COMPUTATIONAL METHODS in LINEAR ALGEBRA

COMPUTATIONAL METHODS in LINEAR ALGEBRA

R. J. Goult
R. F. Hoskins
J. A. Milner
M. J. Pratt

A HALSTED PRESS BOOK

JOHN WILEY & SONS
New York - Toronto

First published in 1974 by Stanley Thornes (Publishers) Ltd.,
17 Quick Street, London N1 8HL.

Published in the U.S.A., Canada
and Latin America by Halsted Press,
a division of John Wiley & Sons, Inc.,
New York

Library of Congress Cataloging in Publication Data
Main entry under title:

Computational methods in linear algebra.

"A Halsted Press book"
1. Algebras, Linear. I. Goult, R.J.
QA184.C65 1975 512'.5 75–19054
ISBN 0–470–31920–8

Printed and bound in Great Britain.

TABLE OF CONTENTS

INTRODUCTION

This book is designed for engineers and scientists who are engaged in work which requires the use of digital computers in problems involving matrices and sets of linear equations. The object is to provide a reasonably comprehensive survey of the numerical methods available, with just as much theory as is needed for their proper use.

We shall be concerned with two problems: solving sets of linear equations (or, what amounts to the same thing, inverting matrices), and finding the eigenvalues (latent roots) and eigenvectors (latent vectors) of matrices

When the matrices concerned are sufficiently small, neither of these problems presents any real difficulty. For example, to solve the equations

$$\begin{aligned} x + y - z &= 0 \\ x - 2y + z &= 1 \\ 2x - y + z &= 3 \end{aligned}$$

we may use, if we wish, Cramer's rule to obtain

$$x = \begin{vmatrix} 0 & 1 & -1 \\ 1 & -2 & 1 \\ 3 & -1 & 1 \end{vmatrix} \div \begin{vmatrix} 1 & 1 & -1 \\ 1 & -2 & 1 \\ 2 & -1 & 1 \end{vmatrix} = (-3) \div (-3) = 1,$$

$$y = \begin{vmatrix} 1 & 0 & -1 \\ 1 & 1 & 1 \\ 2 & 3 & 1 \end{vmatrix} \div \begin{vmatrix} 1 & 1 & -1 \\ 1 & -2 & 1 \\ 2 & -1 & 1 \end{vmatrix} = (-3) \div (-3) = 1,$$

$$z = \begin{vmatrix} 1 & 1 & 0 \\ 1 & -2 & 1 \\ 2 & -1 & 3 \end{vmatrix} \div \begin{vmatrix} 1 & 1 & -1 \\ 1 & -2 & 1 \\ 2 & -1 & 1 \end{vmatrix} = (-6) \div (-3) = 2.$$

The values of these determinants can be written down with almost no intermediate calculations. However, if they were evaluated by a digital computer using an elementary expansion technique, the total number of operations required would be about 39 multiplications and 20 divisions. This, of course, is negligible for a high-speed computer, but even a set of ten equations in ten unknowns (which is an almost trivially small problem by modern standards) would require some 68 million multiplications. This is clearly excessive, and should be compared with the 500 or so required in an elimination process.

Similarly, the extraction of the eigenvalues of a 3 x 3 matrix $[a_{ij}]$ may be accomplished by expanding the function

$$f(\lambda) = \begin{vmatrix} a_{11} - \lambda & a_{12} & a_{13} \\ a_{21} & a_{22} - \lambda & a_{23} \\ a_{31} & a_{32} & a_{33} - \lambda \end{vmatrix}$$

as a polynomial in λ, and then finding its zeros. But even with a third order matrix it is most inadvisable to use an analytic expression for the zeros of the resulting cubic; for matrices of order greater than four, finding eigenvalues necessarily depends upon some numerical approximation method.

In general, expressions which are of importance in theoretical work may contribute little to numerical methods. Further, it is not sufficient to develop methods which appear to be reasonably efficient in terms solely of the numbers of operations required; it is also important to avoid introducing unnecessary errors. In principle, it is possible to solve exactly a set of linear equations, but this is not usually possible in practice. Consider, for example, the case of just three equations, where the coefficients are five decimal digit numbers. We should need to work to fifteen decimal digits to obtain (in all cases) an exact solution. The product of two five-digit numbers requires ten digits for its exact representation, and when this number is subsequently multiplied by another five-digit number, fifteen digits are required.

Even for very moderately sized sets of equations, and for very moderate accuracy in data, the number of digits required for exact evaluation throughout becomes prohibitively large. The usual procedure is to keep constant the number of decimal digits available to represent a number; at each multiplication there is then a "rounding error" introduced. The effects of rounding errors may vary greatly, depending on exactly how an elimination is carried out. However, with large sets of equations, if care is not taken, similar errors can occur even when calculations are carried out correct to nine or ten decimal places. Similar effects may be observed when eigenvalues and eigenvectors are evaluated.

Throughout, we assume little more than a knowledge of elementary matrix algebra and of the basic theory of simultaneous linear equations. Such results, or concepts, of a more advanced nature as are needed are introduced during the course of the text. It was thought advisable, however, to begin with a fairly comprehensive account of the essential properties of eigenvalues and eigenvectors, since this material is rather less likely to be familiar. For the most part, an acquaintance with the more abstract approach of linear algebra proper is not really necessary for an understanding of the text. An introduction to relevant portions of the theory of linear vector spaces, vector and matrix norms, etc. is provided in an Appendix. This is intended partly to amplify some of the discussion in the main text, and partly to serve as a bridge between this and more advanced works in the field.

Included in the text are detailed descriptions of efficient practical algorithms for solving sets of linear equations and computing eigenvalues and eigenvectors.

Largely because of the limitation of the space available, no computer programs have been included. Most computer centres should have library routines which use the methods described. (For example, the NAG library produced by the Nottingham Algorithm Group has a very comprehensive collection of matrix routines both in the FORTRAN and ALGOL languages.) A comprehensive collection of ALGOL programs can be found in *Handbook for Automatic Computation*, Vol. II (Wilkinson *et al*), Springer-Verlag 1971.

Notation Matrices will be denoted by capital letters, A, B, C, etc. Where convenient, we shall use the double suffix notation, a_{ij}, to denote the entry in the ith row and j^{th} column of a matrix A, and write $A \equiv a_{ij}$. The transpose of a matrix A will be denoted by A'; hence, the transpose of an $m \times n$ matrix $A \equiv [a_{ij}]$ is the $n \times m$ matrix A' whose entry in the ith row and j^{th} column is a_{ji}.

Vectors will be identified with *column-vectors* (i.e. $n \times 1$ matrices) and will be distinguished by using Clarendon type, **x**. A *row-vector* (i.e. a $1 \times n$ matrix) will therefore appear as the transpose $\mathbf{x}'$. Thus, the scalar product for a pair of real vectors $\mathbf{x} = (x_1, x_2, \ldots, x_n)$ and $\mathbf{y} = (y_1, y_2, \ldots, y_n)'$ may be variously written as

$$\mathbf{x} \cdot \mathbf{y} = \mathbf{x}'\mathbf{y} = x_1 y_1 + x_2 y_2 + \ldots + x_n y_n$$

In the main we are concerned with real algebra. However, there are occasions when the use of complex matrices is obligatory. (Note, for example, that the eigenvectors of a real matrix generally contain complex entries.) The scalar product for complex vectors takes the form

$$\bar{\mathbf{x}}'\mathbf{y} = \bar{x}_1 y_1 + \bar{x}_2 y_2 + \ldots + \bar{x}_n y_n.$$

In the early part of the text, we sometimes refer to the *length*, $|\mathbf{x}|$, of a vector:

$$|\mathbf{x}| = +\sqrt{\mathbf{x} \cdot \mathbf{x}} = +\sqrt{\{|x_1|^2 + |x_2|^2 + \ldots + |x_n|^2\}}$$

Later, this term is absorbed in the more general concept of *norm*, $\|\mathbf{x}\|$.

The determinant of a square matrix A will usually be denoted by $|A|$. However, when lengthy or complicated expressions are involved, we resort to the less confusing notation

$$\det\{\ldots\}.$$

Special Matrices

A square matrix A is said to be *Hermitian* if $\bar{A}' = \bar{A}$ (where $\bar{A}$ denotes the matrix whose elements are the complex conjugates of those of A). If $\bar{A}' = -A$, then A is said to be *skew-Hermitian*. For real matrices, the terms Hermitian and skew-Hermitian may be replaced by symmetric and skew-symmetric respectively:

$$\text{symmetric} \quad A' = A$$
$$\text{skew-symmetric} \quad A' = -A$$

A matrix is said to be *unitary* if $\bar{A}'A = I$; a real unitary matrix ($A'A = I$) is said to be *orthogonal.*

A *diagonal matrix* is a square matrix whose only non-zero entries lie on the leading diagonal:

$$a_{ii} = \lambda_i, \qquad a_{ij} = 0 \text{ if } i \neq j.$$

We usually write

$$A = \text{diag}\{\lambda_1, \lambda_2, \ldots, \lambda_n\} \equiv \text{diag}\{\lambda_i\}.$$

An *upper triangular matrix* is a square matrix whose only non-zero entries lie on or above the leading diagonal:

$$a_{ij} = 0 \quad \text{if} \quad i > j.$$

Similarly, for a *lower triangular matrix* we have

$$a_{ij} = 0 \quad \text{if} \quad i < j.$$

If $a_{ij} = 0$ for all i and j such that $|i-j| > 1$ then A is said to be a *tridiagonal matrix.*

If $a_{ij} = 0$ for all i and j such that $i > j + 1$ then A is called an *upper Hessenberg matrix.* Similarly, if $a_{ij} = 0$ for all i and j such that $j > i + 1$ then A is called a *lower Hessenberg matrix*.

The so-called *elementary row operations* on a matrix are as follows:

(i) interchange of any two rows,

(ii) multiplication of a row by a non-zero constant,

(iii) addition of a constant multiple of any one row to any other.

An $n \times n$ *elementary matrix* E is a matrix obtained from the $n \times n$ identity matrix I by performing a single elementary row operation upon it. If A is any $n \times m$ matrix the product EA is the matrix obtained by performing the corresponding elementary row operation on A.

CHAPTER 1

EIGENVALUES AND EIGENVECTORS

1.0 Introduction

This chapter is concerned with establishing the more important properties of the eigenvalues and eigenvectors of an $n \times n$ matrix. We shall generally deal with real matrices but, using complex arithmetic, most of the results obtained and the methods discussed are applicable to the complex case.

The problem of computing eigenvalues and eigenvectors is one of considerable practical importance. For example, the natural frequencies of a vibrating mechanical system, or of an electrical network, are determined by the eigenvalues of an associated matrix. Again, in numerical analysis, the convergence of an iterative method for solving a system of linear equations depends on the eigenvalues of a related matrix being sufficiently small. In a later chapter (Chapter 6) the computational problem itself is examined in some detail.

1.1 The Eigenvalue Problem

1.1.1 Basic Definitions

Let A be an $n \times n$ matrix and $\mathbf{v}$ a non-zero column vector such that $A\mathbf{v} = \lambda\mathbf{v}$ for some scalar λ. Then $\mathbf{v}$ is said to be an *eigenvector* of A and λ the related *eigenvalue.* The *eigenvalue problem* for A is the problem of finding all possible pairs λ, $\mathbf{v}$, which satisfy this relation.

It is deceptively simple to produce an equation from which the eigenvalues of A can be computed (in principle, at least). Essentially we have to seek solutions of the matrix equation

$$(A - \lambda I)\mathbf{x} = 0 \tag{1}$$

and a necessary and sufficient condition for this to have a non-trivial solution is given by

$$f(\lambda) \equiv |A - \lambda I| = 0 \tag{2}$$

Equation (2) is a polynomial equation in λ and is called the *characteristic equation* of the matrix A; $f(\lambda)$ itself is called the *characteristic polynomial* of A.

Thus, the problem of finding the eigenvalues of A reduces to the problem of finding the roots of the characteristic polynomial. For matrices of low order, this does provide a practical means of determining eigenvalues, but for matrices of order greater than 4 it is not more practical than the use of Cramer's rule to solve systems of linear equations. For matrices of high order it is difficult to compute the coefficients of $f(\lambda)$ with sufficient accuracy; moreover, in many cases the extraction of the roots of $f(\lambda)$ turns out to be a poorly conditioned problem whereas the eigenvalue problem itself is well-conditioned. In fact, some authors (e.g. Ralston and Wilf: *Mathematical Methods for Digital Computers*) suggest that one approach to the polynomial problem is to construct a matrix of which it is the characteristic polynomial and so convert the problem to an eigenvalue calculation.

Nonetheless, the characteristic polynomial is of considerable theoretical importance. For an $n \times n$ matrix A the characteristic polynomial is of degree n; hence A must have precisely n eigenvalues $\lambda_1, \lambda_2, \ldots, \lambda_n$ (not necessarily all distinct). Again, elementary application of the formula relating roots of a polynomial to its coefficients shows that

$$\lambda_1, \lambda_2 + \ldots + \lambda_n = a_{11} + a_{22} + \ldots + a_{nn} = \text{trace }(A)$$

and,

$$\lambda_1 \lambda_2 \ldots \lambda_n = |A|.$$

The first of these results provides a particularly simple check which at least indicates any gross errors made in a hand calculation.

1.1.2. Worked Examples

(1) Find the characteristic polynomial, the eigenvalues and corresponding eigenvectors of the matrix

$$A = \begin{pmatrix} 1 & 0 & 1 \\ 2 & 0 & 2 \\ 4 & 2 & 3 \end{pmatrix}$$

Solution: The characteristic polynomial is

$$\begin{vmatrix} 1-\lambda & 0 & 1 \\ 2 & -\lambda & 2 \\ 4 & 2 & 3-\lambda \end{vmatrix} \begin{aligned} &= (1-\lambda)(\lambda^2 - 3\lambda - 4) + 4 + 4\lambda \\ &= -\lambda^3 + 4\lambda^2 + 5\lambda \equiv f(\lambda). \end{aligned}$$

In this case $f(\lambda)$ factorises readily:

$$f(\lambda) = -\lambda(\lambda - 5)(\lambda + 1)$$

and so the eigenvalues A are $0, -1$ and 5. The corresponding eigenvectors are obtained by solving, for appropriate values of λ, the equations

$$(A - \lambda I)\mathbf{x} = 0$$

In particular, for $\lambda_1 = 5$ the equations to be solved are

$$\begin{aligned} -4x_1 \quad\quad + x_3 &= 0 \\ 2x_1 - 5x_2 + 2x_3 &= 0 \\ 4x_1 + 2x_2 - 2x_3 &= 0 \end{aligned}$$

One solution of this (singular) system of equations is $x_1 = 1, x_2 = 2, x_3 = 4$, or $\mathbf{v}_1 = (1, 2, 4)'$.

Any scalar multiple of $\mathbf{v}_1$ is also an eigenvector for $\lambda = 5$. Similarly for $\lambda = 0, -1$, the eigenvectors are respectively $\mathbf{v}_2 = (-2, 1, 2)'$ and $\mathbf{v}_3 = (1, 2, -2)'$.

(2) Show that $\lambda = 2$ is an eigenvalue of the matrix

$$A = \begin{pmatrix} 0 & 2 & -6 \\ 1 & -1 & 7 \\ 1 & -1 & 5 \end{pmatrix}.$$

Find the remaining eigenvalues and all the eigenvectors.

Solution: The characteristic polynomial is:

$$\begin{vmatrix} -\lambda & 2 & -6 \\ 1 & -1-\lambda & 7 \\ 1 & -1 & 5-\lambda \end{vmatrix} = (2-\lambda)(\lambda^2 - 2\lambda + 2).$$

In this case, even though the matrix is real, it has a pair of complex eigenvalues. The eigenvalues are $\lambda_1 = 2, \lambda_2 = 1 + i, \lambda_3 = 1 - i$.

Solving $(\mathrm{A} - \lambda I)\mathbf{v} = 0$, as in the previous example, we obtain for $\lambda_1 = 2$ the real eigenvector $\mathbf{v}_1 = (-1, 2, 1)'$.

For the complex eigenvalues $\lambda_2 = 1 + i, \lambda_3 = 1 - i$, the equations to be solved are complex. However, note that $\lambda_3 = \overline{\lambda}_2$. Hence, if $\mathbf{v}_2$ is a solution of $(A - \lambda_2 I)\mathbf{v}_2 = 0$, then taking complex conjugates gives $(A - \overline{\lambda}_2 I)\overline{\mathbf{v}}_2 = 0$. Thus, one value of $\mathbf{v}_3$ is $\mathbf{v}_3 = \overline{\mathbf{v}}_2$. To find a pair of complex eigenvectors, it is thus only necessary to solve the complex linear equations once:

$$\begin{aligned}(-1-i)z_1 + 2z_2 - 6z_3 &= 0\\ z_1 - (2+i)z_2 + 7z_3 &= 0\\ z_1 - z_2 + (4-i)z_3 &= 0, \text{ where } \mathbf{v}_2 = (z_1, z_2, z_3)'\end{aligned}$$

Eliminating z_1 between the last pair of equations gives:

$$z_2 = \frac{3+i}{1+i} z_3 = (2-i)z_3$$

Hence a solution is $z_3 = 1, z_2 = 2-i, z_1 = -2$, and so we get $\mathbf{v}_2 = (-2, 2-i, 1)'$ and $\mathbf{v}_3 = (-2, 2+i, 1)'$.

Note that any complex multiples of $\mathbf{v}_2$ and $\mathbf{v}_3$ will also be eigenvectors. In this way, we can obtain a pair of eigenvectors which are not conjugates and which are not easily recognised as simple multiples of $\mathbf{v}_2$ and $\mathbf{v}_3$. For example, multiplying $\mathbf{v}_2$ by $1+i$, $\mathbf{v}_3$ by i gives the pair of vectors:

$$\mathbf{v}_2^* = (-2-2i, 3+i, 1+i), \qquad \mathbf{v}_3^* = (-2i, -1+2i, i).$$

This property of complex eigenvectors makes it difficult to compare complex eigenvectors found by different numerical algorithms unless the vectors are first scaled by making the element of largest modulus in each vector equal to 1.

1.2 Properties of Eigenvalues and Eigenvectors

1.2.1 Similar Matrices

Square matrices A and B are said to be *similar* if there is a non-singular matrix P such that $B = P^{-1}AP$. In the theory of linear transformations of finite dimensional vector spaces similar matrices play an important role; with a suitable choice of bases similar matrices can represent the same linear transformation. For the algebraic eigenvalue problem similar matrices have important properties which are summarised in the Theorem below.

Theorem

If A is an $n \times n$ matrix, P a non-singular $n \times n$ matrix and $B = P^{-1}AP$, then A and B have the same eigenvalues. If $\mathbf{v}$ is any eigenvector of A then B has a corresponding eigenvector $\mathbf{w} = P^{-1}\mathbf{v}$.

Proof

Let λ be any eigenvalue of A and $\mathbf{v}$ be the corresponding eigenvector. Then,

$$A\mathbf{v} = \lambda\mathbf{v}$$

$$\text{But} \qquad B = P^{-1}AP.$$

Thus, if $\mathbf{w} = P^{-1}\mathbf{v}$, we obtain:

$$\begin{aligned} B\mathbf{w} &= P^{-1}AP(P^{-1}\mathbf{v}) = P^{-1}A\mathbf{v} = P^{-1}(\lambda\mathbf{v}) \\ &= \lambda P^{-1}\mathbf{v} = \lambda\mathbf{w}. \end{aligned}$$

Hence λ is also an eigenvalue of B and $\mathbf{w}$ is the corresponding eigenvector. Conversely, we can easily prove that if μ is an eigenvalue of B, with corresponding eigenvector $\mathbf{w}$, then μ is an eigenvalue of A corresponding to $\mathbf{v} = P\mathbf{w}$.

Note: By considering pairs of non-symmetric matrices with repeated eigenvalues, it is easy to find counter-examples showing that having the same eigenvalues is not sufficient to prove similarity for a pair of matrices (see Exercise 9, p. 24.

This theorem provides the essential theoretical justification for many practical algorithms for computing eigenvalues and eigenvectors. These algorithms reduce the problem of finding the eigenvalues of a given matrix A to that of finding the eigenvalues of another matrix $B = P^{-1}AP$. A suitable choice of P ensures that B has a simpler structure than A. For example in the Givens–Householder method described in detail in the next chapter, A is a symmetric matrix and P is chosen so that B is a tridiagonal matrix.

1.2.2 Theorem

If A has distinct eigenvalues $\lambda_1, \lambda_2, \ldots, \lambda_m$ then the corresponding eigenvectors $\mathbf{v}_1, \mathbf{v}_2, \ldots, \mathbf{v}_m$ are linearly independent.

Proof

Suppose, if possible, that the eigenvectors $\mathbf{v}_1, \mathbf{v}_2, \ldots, \mathbf{v}_m$ are linearly dependent. Without loss of generality we may assume that $\mathbf{v}_1, \mathbf{v}_2, \ldots, \mathbf{v}_k$ are linearly independent while $\mathbf{v}_1, \ldots, \mathbf{v}_k, \mathbf{v}_{k+1}$ form a linearly dependent set. Then, for some constants $\alpha_1, \alpha_2, \ldots, \alpha_k$, we must have

$$\mathbf{v}_{k+1} = \sum_{i=1}^{k} \alpha_i \mathbf{v}_i \tag{3}$$

Multiplying equation (3) by matrix A gives:

$$\lambda_{k+1}\mathbf{v}_{k+1} = A\mathbf{v}_{k+1} = \sum_{i=1}^{k} \alpha_i A\mathbf{v}_i = \sum_{i=1}^{k} \alpha_i \lambda_i \mathbf{v}_i \tag{4}$$

Multiplying equation (3) by λ_{k+1} gives:

$$\lambda_{k+1}\mathbf{v}_{k+1} = \sum_{i=1}^{k} \alpha_i \lambda_{k+1} \mathbf{v}_i \tag{5}$$

Subtracting (4) from (5) shows that

$$\sum_{i=1}^{k} \alpha_i(\lambda_{k+1} - \lambda_i)\mathbf{v}_i = 0.$$

Since $\lambda_{k+1} - \lambda_i \neq 0$, for any $i = 1, \ldots, k$ and $\mathbf{v}_1, \mathbf{v}_2, \ldots, \mathbf{v}_k$ are independent, we must have $\alpha_i = 0$ for all i. Hence $\mathbf{v}_{k+1} = 0$ which is impossible.

Corollary

If an $n \times n$ matrix A has a distinct eigenvalues $\lambda_1, \lambda_2, \ldots, \lambda_n$ then A is similar to the diagonal matrix $\operatorname{diag}(\lambda_1, \ldots, \lambda_n)$.

Proof

Since A has distinct eigenvalues, it has a set $\mathbf{v}_1, \mathbf{v}_2, \ldots, \mathbf{v}_n$ of linearly independent eigenvectors. Let P be the matrix whose column vectors are $\mathbf{v}_1, \mathbf{v}_2, \ldots, \mathbf{v}_n$; i.e. $P = (\mathbf{v}_1\, \mathbf{v}_2 \ldots \mathbf{v}_n)$. Since the columns of P are linearly independent, P is a non-singular matrix and has an inverse P^{-1}.

Consider $P^{-1}AP$:

$$P^{-1}AP = P^{-1}A(\mathbf{v}_1, \mathbf{v}_2, \ldots, \mathbf{v}_n) = P^{-1}(A\mathbf{v}_1, A\mathbf{v}_2, \ldots, A\mathbf{v}_n)$$

But, for each i, $A\mathbf{v}_i = \lambda_i \mathbf{v}_i$

Hence,
$$\begin{aligned} P^{-1}AP &= P^{-1}(\lambda_1\mathbf{v}_1, \lambda_2\mathbf{v}_2, \ldots, \lambda_n\mathbf{v}_n). \\ &= P^{-1}(\mathbf{v}_1, \mathbf{v}_2, \ldots, \mathbf{v}_n)\operatorname{diag}(\lambda_1, \ldots, \lambda_n) \\ &= \operatorname{diag}(\lambda_1, \ldots, \lambda_n). \end{aligned}$$

Notes

(i) The success of the above method of reducing A to diagonal form by a similarity transformation depends essentially upon the fact that A has a complete eigensystem consisting of n linearly independent eigenvectors. For matrices with this property, the reduction is always possible, even if A has repeated eigenvalues.

(ii) The above reduction has important applications to the solution of systems of linear differential equations (see Example 2, p. 12 and Exercise 11, p. 24).

(iii) In vector space theory, the reduction of A to diagonal form is equivalent to expressing the related linear transformation in terms of a basis which consists entirely of eigenvectors.

1.2.3 Theorem

The matrices A and A' always have the same eigenvalues.

Proof

For any square matrix M, we have $|M| = |M'|$. Hence

$$|A - \lambda I| = |(A - \lambda I)'| = |A' - \lambda I|$$

so that A and A' have the same characteristic equation and therefore the same eigenvalues.

Although A and A' have the same eigenvalues they will not generally have the same eigenvectors. For any eigenvalue λ, the corresponding eigenvector of A is a solution of:

$$A\mathbf{x} = \lambda\mathbf{x} \tag{6}$$

whereas the eigenvector of A' is a solution of:

$$A'\mathbf{y} = \lambda\mathbf{y} \tag{7}$$

The eigenvectors of A' are sometimes referred to as the *left-hand eigenvectors* of A, since transposing equation (7) gives $\mathbf{y}'A = \lambda\mathbf{y}'$.

As will be seen in a subsequent chapter, the left-hand eigenvectors of a matrix play a significant part in determining whether or not its eigenvalue problem is well conditioned.

1.2.4 Worked Examples

1. Find the left- and right-hand eigenvectors of the matrix

$$A = \begin{pmatrix} -1 & 1 & -1 \\ -6 & 5 & 0 \\ 6 & -5 & -2 \end{pmatrix}.$$

Show that $\mathbf{w}_i'\mathbf{v}_j = 0$, for $i \neq j$, where $\mathbf{v}_j$ denotes the right eigenvector corresponding to λ_j, and $\mathbf{w}_i$ denotes the left eigenvector corresponding to λ_i.

Solution: The characteristic polynomial of A is

$$\begin{vmatrix} -1-\lambda & 1 & -1 \\ -6 & 5-\lambda & 0 \\ 6 & -5 & -2-\lambda \end{vmatrix} = (-1-\lambda)(\lambda^2 - 3\lambda - 2)$$

so that the eigenvalues are $-1, 1, 2$.

Using the notation of the question, $\mathbf{v}_1$ is given by solving $(A+I)\mathbf{v}_1 = 0$ and $\mathbf{w}_1$ by solving $(A'+I)\mathbf{w}_1 = 0$:

$$\begin{pmatrix} 0 & 1 & -1 \\ -6 & 6 & 0 \\ 6 & -5 & -1 \end{pmatrix}\mathbf{v}_1 = 0 \qquad \text{and} \qquad \begin{pmatrix} 0 & -6 & 6 \\ 1 & 6 & -5 \\ -1 & 0 & -1 \end{pmatrix}\mathbf{w}_1 = 0$$

whence, $\mathbf{v}_1 = (1, 1, 1)'$ and $\mathbf{w}_1 = (-1, 1, 1)'$.

Similarly by solving the appropriate linear equations.

$$\mathbf{v}_2 = (2, 3, -1), \qquad \mathbf{w}_2 = (3, 2, -1)$$
$$\mathbf{v}_3 = (1, 2, -1), \qquad \mathbf{w}_3 = (4, -3, -1).$$

It can now be quickly verified that

$$\mathbf{w}_1'\mathbf{v}_2 = \mathbf{w}_1'\mathbf{v}_3 = \mathbf{w}_2'\mathbf{v}_3 = 0, \text{ etc.}$$

2. A system of linear differential equations is expressible in matrix form as $\dot{\mathbf{x}} = A\mathbf{x}$, where A is the matrix given in Example 1. Show that, by means of a suitable change of variables, $\mathbf{x} = P\mathbf{y}$, the equations can be expressed in an "uncoupled" form $\dot{\mathbf{y}} = D\mathbf{y}$, where D is a diagonal matrix. Hence find the general solution.

Solution: The matrix A has a complete eigensystem $\mathbf{v}_1, \mathbf{v}_2, \mathbf{v}_3$. Thus if

$$P = \begin{pmatrix} 1 & 2 & 1 \\ 1 & 3 & 2 \\ 1 & -1 & -1 \end{pmatrix}$$

we have $P^{-1}AP = \text{diag}(-1, 1, 2)$. Making the substitution $\mathbf{x} = P\mathbf{y}$ reduces the system of differential equations to:

$$\dot{\mathbf{x}} = P\dot{\mathbf{y}} = AP\mathbf{y}$$

$$\text{or} \quad \dot{\mathbf{y}} = P^{-1}AP\mathbf{y} = \text{diag}(-1, 1, 2)\mathbf{y} \tag{8}$$

Written out fully, equations (8) are

$$\frac{dy_1}{dt} = -y_1$$

$$\frac{dy_2}{dt} = y_2$$

$$\frac{dy_3}{dt} = 2y_3.$$

These can now be solved individually as first order linear differential equations to give $y_1 = \alpha_1 e^{-t}$, $y_2 = \alpha_2 e^{t}$, $y_3 = \alpha_3 e^{2t}$, where the α_i are arbitrary constants.

Since $\mathbf{x} = P\mathbf{y}$ the general solution of the original system of differential equations is thus:

$$\mathbf{x} = \alpha_1 e^{-t}\begin{pmatrix} 1 \\ 1 \\ 1 \end{pmatrix} + \alpha_2 e^{t}\begin{pmatrix} 2 \\ 3 \\ -1 \end{pmatrix} + \alpha_3 e^{2t}\begin{pmatrix} 1 \\ 2 \\ -1 \end{pmatrix}.$$

Note that the exponential indices occurring in this expression are the eigenvalues of A and the appropriate coefficient vectors are the eigenvectors.

1.3 Properties of Real Symmetric Matrices

1.3.1

Real symmetric matrices (and indeed their complex generalisation, Hermitian matrices) have a number of important properties which can be exploited when solving eigenvalue problems. Such matrices always have real eigenvalues, always have a complete eigensystem and always have a well-conditioned eigenvalue problem. Even apart from the saving in computer storage space made available by exploiting the symmetry, these properties are of sufficient importance to make the development of special algorithms for dealing with the symmetric eigenproblem well worth while.

1.3.2 Theorem

The eigenvalues of a real symmetric matrix are all real.

Proof

Let λ be an eigenvalue and $\mathbf{v}$ an associated eigenvector of the symmetric matrix A. Since λ might be complex, we should assume that, in general, $\mathbf{v} = \mathbf{x} + i\mathbf{y}$ where $\mathbf{x}$ and $\mathbf{y}$ are real vectors.

Now, $$A\mathbf{v} = \lambda \mathbf{v} \tag{9}$$

and so $$\bar{\mathbf{v}}'A\mathbf{v} = \lambda \bar{\mathbf{v}}'\mathbf{v} \tag{10}$$

$$\bar{\mathbf{v}}'\mathbf{v} = (\bar{v}_1, \bar{v}_2, \ldots, \bar{v}_n)(v_1, v_2, \ldots, v_n)'$$

$$= \sum_{i=1}^{n} |v_i|^2, \text{ which is real and positive.}$$

Taking the complex conjugate of equation (9), transposing, and post-multiplying by $\mathbf{v}$ gives:

$$(\bar{\mathbf{v}}'\bar{A}')\mathbf{v} = \bar{\lambda}\bar{\mathbf{v}}'\mathbf{v}$$

But, since A is real and symmetric, $\bar{A}' = A$, and so

$$\bar{\mathbf{v}}'A\mathbf{v} = \bar{\lambda}\bar{\mathbf{v}}'\mathbf{v} \tag{11}$$

Comparing equations (10) and (11) shows that

$$\lambda\bar{\mathbf{v}}'\mathbf{v} = \bar{\lambda}\bar{\mathbf{v}}'\mathbf{v}.$$

Since $\bar{\mathbf{v}}'\mathbf{v} \neq 0$ we obtain immediately $\lambda = \bar{\lambda}$.

Note: Since all the eigenvalues of a symmetric matrix are real it follows immediately that the eigenvectors are also real.

1.3.3 Theorem (Orthogonal Property)

If $\mathbf{v}_1$ and $\mathbf{v}_2$ are eigenvectors of a real symmetric matrix A, corresponding to distinct eigenvalues λ_1 and λ_2, then $\mathbf{v}_1 . \mathbf{v}_2 = \mathbf{v}_1'\mathbf{v}_2 = 0$.

Proof

$$A\mathbf{v}_1 = \lambda_1\mathbf{v}_1 \qquad \text{and} \qquad A\mathbf{v}_2 = \lambda_2\mathbf{v}_2$$

Hence, by transposing $\mathbf{v}_1'A = \lambda_1 \mathbf{v}_1'$

Thus $$(\mathbf{v}_1'A)\mathbf{v}_2 = \lambda_1 \mathbf{v}_1'\mathbf{v}_2 = \mathbf{v}_1'(A\mathbf{v}_2) = \lambda_2 \mathbf{v}'_1\mathbf{v}_2,$$

and so $$(\lambda_1 - \lambda_2)\mathbf{v}_1'\mathbf{v}_2 = 0.$$

Since $\lambda_1 \neq \lambda_2$ we must have $\mathbf{v}_1'\mathbf{v}_2 = 0$, i.e. $\mathbf{v}_1 . \mathbf{v}_2 = 0$.
That is to say, the eigenvectors of a symmetric matrix corresponding to distinct eigenvalues are orthogonal.

1.3.4 Theorem (Diagonalisation Theorem)

If A is a real symmetric matrix then there exists a real orthogonal matrix P such that P'AP = D is a diagonal matrix.

Proof

Any matrix A is similar to a triangular matrix (see, for example, Wilkinson, [1] p. 46). That is to say, we know there exists a matrix P such that $P'AP = T$ is a triangular matrix. Moreover, since the eigenvalues of A are all real, the column vectors of P are real. But, since A is symmetric, $T = T'$. Hence T is a symmetric triangular matrix and must therefore be diagonal.

Note: The diagonal elements of D are of course the eigenvalues of A and the column vectors of P form a set of mutually orthogonal eigenvectors for A. An obvious implication of this theorem is that when a symmetric matrix A has an eigenvalue λ which is r times repeated then it always has a set of r mutually orthogonal (and hence linearly independent) eigenvectors corresponding to λ.

A further corollary to the above theorems is that any two symmetric matrices with the same eigenvalues are similar.

1.3.5 Quadratic Forms

Each real symmetric matrix A can be associated with a *quadratic form* $\phi(x_1, x_2, \ldots, x_n) = \mathbf{x}'A\mathbf{x}$. For example the matrix

$$A = \begin{pmatrix} 2 & 1 & -1 \\ 1 & 3 & 2 \\ -1 & 2 & 5 \end{pmatrix}$$

is associated with the quadratic expression

$$\phi = 2x_1^2 + 3x_2^2 + 5x_3^2 + 2x_1x_2 - 2x_3x_1 + 4x_2x_3.$$

A quadratic form ϕ is said to be *positive definite* if it is positive for every set of (real) values of the variables $x_1, x_2, \ldots, x_n$, other than the set $x_1 = x_2 = \ldots = x_n = 0$.

A matrix A is said to be positive definite if its associated quadratic form is positive definite. The linear substitution $\mathbf{x} = P\mathbf{y}$, where P is an orthogonal matrix, reduces ϕ to the quadratic form $\mathbf{y}'(P'AP)\mathbf{y}$ in the new variables $y_1, y_2, \ldots, y_n$. A particular choice of P is that which reduces A to the diagonal form; in this case $\phi(y_1, \ldots, y_n)$ has the particularly simple form: $\lambda_1 y_1^2 + \lambda_2 y_2^2 + \ldots + \lambda_n y_n^2$. From this, we can deduce the fact that A is positive definite if and only if its eigenvalues are all positive.

Positive definite matrices have a number of important applications. Even the comparatively simple problem of solving a set of linear equations can be further simplified if it is known that the coefficient matrix is positive definite (see Section 3.2.6). The test given above is only one of a number of ways of determining whether or not a given symmetric matrix is positive definite. For further criteria see, for example, Exercise 5, p. 23, or any advanced linear algebra textbook.

More generally, if A is a Hermitian matrix then we may associate with it a so-called *Hermitian form*, $\bar{\mathbf{x}}'A\mathbf{x}$.

Note that $$\overline{(\bar{\mathbf{x}}'A\mathbf{x})'} = \bar{\mathbf{x}}'\bar{A}'\mathbf{x}$$

so that $\bar{\mathbf{x}}'A\mathbf{x}$ is actually always real. The above remarks on positive definiteness generalise in an obvious way to the case of Hermitian forms and Hermitian matrices.

1.4 The Power Method

1.4.1

The so-called *power method* is a simple iterative method of calculating eigenvalues and eigenvectors. Apart from the case of large, sparse matrices, it is of limited practical importance, more sophisticated methods being usually preferable. However, it is of some historical interest as the first indirect method of performing the calculation. Moreover, it serves as a useful illustration of the principles underlying all iterative methods, including the widely used Q–R method.

1.4.2 Theory of the Method

For simplicity, we assume that A is a real, $n \times n$ matrix with n real, distinct, eigenvalues $\lambda_1, \lambda_2, \ldots, \lambda_n$. A thus has a full set of n linearly independent eigenvectors $\mathbf{v}_1, \ldots, \mathbf{v}_n$. These form a *basis* for the space, R^n, of all n-dimensional column vectors in the sense that any such vector $\mathbf{x}$ admits a representation of the form,

$$\mathbf{x} = \alpha_1 \mathbf{v}_1 + \alpha_2 \mathbf{v}_2 + \ldots + \alpha_n \mathbf{v}_n.$$

We shall also assume that the eigenvalues are ordered so that

$$|\lambda_1| \geqslant |\lambda_2| \geqslant \ldots \geqslant |\lambda_n|.$$

Now for each i,

$$A\mathbf{v}_i = \lambda_i \mathbf{v}_i$$

so that

$$A\mathbf{x} = A\left\{\sum_{i=1}^{n} \alpha_i \mathbf{v}_i\right\} = \alpha_1 \lambda_1 \mathbf{v}_1 + \alpha_2 \lambda_2 \mathbf{v}_2 + \ldots + \alpha_n \lambda_n \mathbf{v}_n.$$

More generally, for any positive integer k,

$$A^k \mathbf{x} = \alpha_1 \lambda_1^k \mathbf{v}_1 + \alpha_2 \lambda_2^k \mathbf{v}_2 + \ldots + \alpha_n \lambda_n^k \mathbf{v}_n.$$

If we now assume that $\alpha_1 \neq 0$ and that $|\lambda_1| > |\lambda_2|$ we obtain $|\lambda_1^k| \gg |\lambda_i^k|$ for all $i \geqslant 2$ and all sufficiently large k.

The expression for $A^k\mathbf{x}$ is now dominated by $\alpha_1 \lambda_1^k \mathbf{v}_1$, and so $A^k\mathbf{x}$ provides a good approximation for a multiple of $\mathbf{v}_1$ – the eigenvector of A corresponding to the eigenvalue of largest modulus.

The obvious disadvantage is that if $|\lambda_1|$ is large then after a few iterations the elements of $A^k\mathbf{x}$ will be very large, and overflow may soon occur. This can be overcome by normalising this vector after each iteration. Thus, suppose at each iteration we divide by λ_1 as well as multiplying by A. Then the vector produced after k iterations will be

$$\frac{1}{\lambda_1^k} A^k \mathbf{x} = \alpha_1 \mathbf{v}_1 + \sum_{i=2}^{n} \alpha_i \left(\frac{\lambda_i}{\lambda_1}\right)^k \mathbf{v}_i.$$

Clearly, the rate of convergence of the method is now proportional to the rate at which $\left|\frac{\lambda_2}{\lambda_1}\right|^k \to 0$. Convergence will be very slow for matrices having two large eigenvalues of nearly equal modulus. For example, if A is a 3 x 3 matrix with eigenvalues 2.0, −1.9, 0.5, we could expect approximately 50 iterations to be needed to determine $\mathbf{v}_1$ to only 1 significant digit accuracy.

It would also appear that the method would fail to produce the eigenvalue of largest modulus if the initial vector $\mathbf{x}$ were such that $\alpha_1 = 0$. In practice, if $\mathbf{x}$ is chosen arbitrarily, there is little likelihood of this occurring. Even when it does happen, rounding errors, after one or two iterations, will normally introduce a non-zero component of $\mathbf{v}_1$ into the computation. The effect of a poor initial choice of $\mathbf{x}$ is thus to slightly increase the number of iterations needed.

1.4.3 A Simple Algorithm

In the theoretical account of the method, it was asserted that after each iteration the vector produced was divided by λ_1. In practice, of course, λ_1 is not known until the end of the process. However, all that is really necessary is to divide by something which ensures that the possibility of overflow is avoided. In the algorithm below, we always divide by the element of largest modulus in the vector produced.

Algorithm: Let $\mathbf{x}_0$ be the initial vector (say $\mathbf{x}_0 = (1, 1, \ldots, 1)'$).
Let $\mathbf{y}_1 = A\mathbf{x}_0$, and let k_1 be the element of largest modulus in $\mathbf{y}_1$.
Define $\mathbf{x}_1 = \dfrac{1}{k_1}\mathbf{y}_1$.

The iteration now proceeds with

$$\begin{aligned} \mathbf{y}_{r+1} &= A\mathbf{x}_r \\ k_{r+1} &= \text{element of largest modulus in } \mathbf{y}_{r+1} \\ \mathbf{x}_{r+1} &= \frac{1}{k_{r+1}}\mathbf{x}_r. \end{aligned}$$

After each iteration $\mathbf{x}_r$ and $\mathbf{x}_{r+1}$ are compared and the process is terminated when these vectors are sufficiently close.

Provided the problem is well-conditioned (see Chapter 2), the process should result in $\mathbf{x}_{r+1}$ as a good approximation to $\mathbf{v}_1$, and k_{r+1} as a good approximation to λ_1.

1.4.4 Origin Shifts

If A has eigenvalues $\lambda_1, \lambda_2, \ldots, \lambda_n$ then, as can be quickly verified, $A - cI$ has eigenvalues $\lambda_1 - c, \lambda_2 - c, \ldots, \lambda_n - c$ associated with the eigenvectors of A.

We have seen that the success of the simple iterative method depends upon $\left|\dfrac{\lambda_2}{\lambda_1}\right|$ being very small. If, instead of iterating with the matrix A, we use $A - cI$, then the rate of convergence will be enhanced provided that $\max\limits_i \left|\dfrac{\lambda_i - c}{\lambda_1 - c}\right| < \left|\dfrac{\lambda_2}{\lambda_1}\right|$.
Such a change of origin is particularly beneficial if the two eigenvalues of largest modulus differ in sign. If, for example, A has eigenvalues of 2.0, −1.9, 0.5, iteration with $A + 2I$ enables us to find $\mathbf{v}_1$ at a rate which is dependent upon the rate at which $\left(\dfrac{2.5}{4.0}\right)^k$ tends to 0. Similarly, iteration with $A - 2I$ enables us to find $\mathbf{v}_2$ at a rate which depends upon the rate at which $\left(\dfrac{1.5}{3.9}\right)^k$ tends to 0. Clearly, with suitable origin shifts, the power method can be used to find the largest and the smallest eigenvalues of a matrix with real eigenvalues. If more eigenvalues are needed, the power method must be used in conjunction with some method of *deflation*; that is, some method of modifying the original matrix so that the eigenvalue of largest modulus no longer occurs.

1.4.5 Deflation of a Matrix

To illustrate the principles of deflation methods, we first describe a simple deflation technique which could be applied to symmetric matrices.

Suppose A is a symmetric matrix whose eigenvalues $\lambda_1, \lambda_2, \ldots, \lambda_n$ are arranged in descending order of modulus. Assume $\mathbf{v}_1$ and λ_1 have been determined, with $|\mathbf{v}_1| = 1$. Define a new symmetric matrix, $B = A - \lambda_1 \mathbf{v}_1 \mathbf{v}_1'$. If the eigenvectors of A are $\mathbf{v}_1, \mathbf{v}_2, \ldots, \mathbf{v}_n$ we have

$$\begin{aligned} B\mathbf{v}_1 &= (A - \lambda_1 \mathbf{v}_1 \mathbf{v}_1')\mathbf{v}_1 = A\mathbf{v}_1 - \lambda_1 \mathbf{v}_1 (\mathbf{v}_1' \mathbf{v}_1) \\ &= 0, \text{ since } \mathbf{v}_1'\mathbf{v}_1 = 1 \text{ and } A\mathbf{v}_1 = \lambda_1 \mathbf{v}_1. \end{aligned}$$

Also, for each $i = 2, \ldots, n$:

$$\begin{aligned} B\mathbf{v}_i &= (A - \lambda_1 \mathbf{v}_1 \mathbf{v}_1')\mathbf{v}_i = A\mathbf{v}_i - \lambda_1 \mathbf{v}_1 (\mathbf{v}_1' \mathbf{v}_i) \\ &= A\mathbf{v}_i - 0 \end{aligned}$$

since the eigenvectors are orthogonal (see Section 1.3.3). The matrix B is thus seen to have eigenvectors $\mathbf{v}_1, \mathbf{v}_2, \ldots, \mathbf{v}_n$ associated with the eigenvalues $0, \lambda_2, \lambda_3, \ldots, \lambda_n$. If, after finding λ_1 and $\mathbf{v}_1$, the power method is applied to the deflated matrix B the iterative process should converge to produce λ_2 and $\mathbf{v}_2$. It is clear that we must now define a new matrix $C = B - \lambda_2 \mathbf{v}_2 \mathbf{v}_2'$ in order to find λ_3 and $\mathbf{v}_3$. In theory, at least, this deflation technique can be successively applied until eventually all the eigenvalues and eigenvectors have been found. In practice, this method of deflation is considered to be unreliable since at each stage the eigenvalues and eigenvectors determined are subject to error, and these errors have a cumulative effect on the deflated matrices. A further disadvantage, when compared with other methods of deflation, is that the orders of the deflated matrices are not reduced.

1.4.6 Deflation with Order Reduction

Suppose now that A is a real matrix, not necessarily symmetric, and that the real eigenvalue λ_1 of largest modulus and its associated eigenvector $\mathbf{v}_1$ have been determined. We may assume that $\mathbf{v}_1$ has 1 as its element of largest modulus. If this element is not in the first position, we can easily achieve this by multiplying $\mathbf{v}_1$ by a permutation matrix P which interchanges the largest element and the 1st element (Note that $P = P^{-1}$). Suppose $P\mathbf{v}_1 = \mathbf{w}_1$. It is now necessary to find an elementary matrix R such that $R\mathbf{w}_1 = \mathbf{e}_1$, the elementary vector with first element 1 and all other elements 0.

Let B be the matrix $B = RPAP^{-1}R^{-1} = RPAPR^{-1}$

Then
$$\begin{aligned} B\mathbf{e}_1 &= RPAPR^{-1}\mathbf{e}_1 \\ &= RPAP\mathbf{w}_1 \\ &= RPA\mathbf{v}_1 \\ &= \lambda_1 RP\mathbf{v}_1 = \lambda_1 \mathbf{e}_1. \end{aligned}$$

This shows that B has $\mathbf{e}_1$ as the eigenvector associated with the eigenvalue λ_1. Also, since B is similar to A, its other eigenvalues are identical with those of A and its eigenvectors are all of the form $RP\mathbf{v}_i$. The fact that B has $\mathbf{e}_1$ as an eigenvector means that it must be a matrix of a particularly simple form, namely:

$$B = \left(\begin{array}{c|c} \lambda_1 & \mathbf{x}' \\ \hline 0 & \\ \cdot & \\ \cdot & B_1 \\ \cdot & \\ 0 & \end{array}\right) \tag{12}$$

where B_1 is an $(n-1) \times (n-1)$ matrix and $\mathbf{x}$ is an $(n-1)$ vector which may be non-zero. Since B has the same eigenvalues as A, the eigenvalues $\lambda_2, \lambda_3, \ldots, \lambda_n$ must be precisely the eigenvalues of the smaller, deflated, matrix B_1 which will be used for the next iterations by the power method.

Before this deflation process can be used in a practical algorithm, it is necessary to study two points in some detail:

(a) Given $A, \lambda_1, \mathbf{v}_1$, is there a simple way of calculating B and hence B_1?
(b) Although B_1 will have the correct eigenvalues, its eigenvectors will not be the same as those of A. Some method of recovering the eigenvectors of A from those of B_1 is therefore required.

Considering these two points in order, we have:

(a) Rather than finding explicit expressions for the matrices P and R, it is very much simpler to compute RPA by simply applying the appropriate row interchanges and elementary row operations to the matrix A. $B = RPAP^{-1}R^{-1}$ can then be computed by applying the operations corresponding to P^{-1} and R^{-1} *to* the columns of RPA (the matrix produced at the previous stage). In practice, the operations to be performed on the matrix currently stored are:

(i) Interchange rows 1 and i (assuming that the eigenvector $\mathbf{v}_1$ has 1 as its largest element in row i).
(ii) For $j = 2, \ldots, n$, subtract w_j times row 1 from row j (where $\mathbf{w}$ is the vector obtained from $\mathbf{v}_1$ by interchanging rows 1 and i).
(iii) Interchange columns 1 and i.
(iv) Add w_j times column j to column 1 for $j = 2, \ldots, n$.

Notes

Since for all j we have $|w_j| \leqslant 1$ these operations are numerically stable.

Operations (iv) need not be performed since after the deflation has been completed we know that the first column should be $(\lambda_1, 0, \ldots, 0)'$. Full information about these operations can be retained by storing $\mathbf{v}_1$.

(b) Assume this deflation has been carried out and that the power method has been applied to find the eigenvalue λ_2 and corresponding eigenvector $\mathbf{y}_2$ of the deflated sub-matrix B_1. $\mathbf{y}_2$ will of course be a vector with $n-1$ elements since B_1 is a matrix of order $n-1$. The corresponding eigenvector of the $n \times n$ matrix B of equation (12) is of the form $\mathbf{z} = \begin{pmatrix} z \\ \mathbf{y}_2 \end{pmatrix}$. The element z is determined by the defining condition $B\mathbf{z} = \lambda_2 \mathbf{z}$. Hence, $(\lambda_2 - \lambda_1)z = \mathbf{x}'\mathbf{y}_2$.

Once $\mathbf{z}$ has been determined, the corresponding eigenvector $\mathbf{v}_2$ of the matrix $A = R^{-1}P^{-1}BPR$ is $\mathbf{v}_2 = P^{-1}R^{-1}\mathbf{z}$. In practice, $\mathbf{v}_2$ is computed from $\mathbf{z}$ by applying the inverse operations to (a)(ii) and (a)(i).

Note on Storage Requirements

From the above description, it can be seen that the only information necessary to perform the deflation and to compute eigenvectors of the original matrix from those of the deflated matrix is λ_1, λ_2, $\mathbf{v}_1$, and $\mathbf{x}$. If the calculation is performed on a digital computer and, at each stage, the original matrix is overwritten with the deflated matrix, then the eigenvalues are stored as diagonal elements, the appropriate vectors $\mathbf{x}$ being stored in the rows adjacent to the diagonal elements. Consequently, the only additional storage requirement is for the eigenvectors as they are produced. The entire algorithm is thus very economical in its use of computer storage. This may make it worthy of consideration when only a few eigenvalues of a large matrix are to be computed.

1.4.7 Worked Example

Find the eigenvalues and eigenvectors of the matrix

$$A = \begin{pmatrix} 0 & 5 & -6 \\ -4 & 12 & -12 \\ -2 & -2 & 10 \end{pmatrix},$$

using the power method together with deflation.

The solution, which is intended to illustrate the power method as described above, will use the same notation.

Starting with $\mathbf{x}_0 = (1, 1, 1)'$ produces as successive values of $\mathbf{x}_i$ and k_i the values: (working to 4 significant figures)

k_i	6.0	−19.33	17.31	16.25	16.05	16.01	16.00
$\mathbf{x}_i$	−0.1667	0.4828	0.4981	0.4998	0.4999	0.50	0.50
	−0.6667	1.0	1.0	1.0	1.0	1.0	1.0
	1.0	−0.6034	−0.5191	−0.5044	−0.5010	−0.5003	−0.50

The largest eigenvalue is thus $\lambda_1 = 16$ and this is associated with the eigenvector $\mathbf{v}_1 = (0.5, 1.0, -0.5)'$.

From $\mathbf{v}_1$ the appropriate deflation process is:

(i) interchange rows 1 and 2,
(ii) subtract $\frac{1}{2}$ row 1 from row 2; add $\frac{1}{2}$ row 1 to row 3,
(iii) interchange columns 1 and 2.

The resulting matrices are:

$$\begin{pmatrix} 0 & 5 & -6 \\ -4 & 12 & -12 \\ -2 & -2 & 10 \end{pmatrix} \xrightarrow{\text{(i)}} \begin{pmatrix} -4 & 12 & -12 \\ 0 & 5 & -6 \\ -2 & -2 & 10 \end{pmatrix} \xrightarrow{\text{(ii)}} \begin{pmatrix} -4 & 12 & -12 \\ 2 & -1 & 0 \\ -4 & 4 & 4 \end{pmatrix} \xrightarrow{\text{(iii)}} \begin{pmatrix} 12 & -4 & -12 \\ -1 & 2 & 0 \\ 4 & -4 & 4 \end{pmatrix}$$

Note that if we carry out the final operations of adding the appropriate multiples of the columns to the 1st column we obtain,

$$\begin{pmatrix} 16 & -4 & -12 \\ 0 & 2 & 0 \\ 0 & -4 & 4 \end{pmatrix} = \left(\begin{array}{c|c} \lambda_1 & \mathbf{x}' \\ \hline 0 & \\ 0 & B_1 \end{array}\right)$$

The matrix to be used in the next series of iterations is thus $B_1 = \begin{pmatrix} 2 & 0 \\ -4 & 4 \end{pmatrix}$.

Using the power method again and iterating with matrix B_1 produces the eigenvalue $\lambda_2 = 4$ and associated eigenvector $\mathbf{v}_2 = (0, 1)'$ of B_1. Applying the standard deflation process of matrix B_1 gives the matrix $C = \begin{pmatrix} 4 & -4 \\ 0 & 2 \end{pmatrix}$.

The deflated 1 x 1 submatrix is thus $C_1 = (2)$ which has eigenvalue $\lambda_3 = 2$ associated with the eigenvector $(1)'$. The remaining problem is that of reconstructing from the eigenvectors of the deflated matrices the corresponding eigenvectors of A. B_1 has eigenvector $\mathbf{y}_2 = (0, 1)'$ associated with $\lambda_2 = 4$.

The corresponding eigenvector of

$$B = \begin{pmatrix} 16 & -4 & -12 \\ 0 & 2 & 0 \\ 0 & -4 & 4 \end{pmatrix} \quad \text{is then} \quad \mathbf{z} = \begin{pmatrix} z \\ 0 \\ 1 \end{pmatrix},$$

where $(\lambda_2 - \lambda_1) z = \mathbf{x}'\mathbf{y}_2$ (using the notation of Section 1.4.6, p. 20)

or $$-12z = (-4, -12)(0, 1)'.$$

Thus $$z = 1.$$

The required eigenvector of B is thus $(1,0,1)'$. The corresponding eigenvector for A is given by applying to this vector the operations inverse to (ii) and (i), that is:

(ii)′ add $\frac{1}{2}$ row 1 to row 2; subtract $\frac{1}{2}$ row 1 from row 3,
(i)′ interchange rows 1 and 2,

The results are

$$\begin{pmatrix}1\\0\\1\end{pmatrix} \overset{(ii)'}{\rightarrow} \begin{pmatrix}1\\\frac{1}{2}\\\frac{1}{2}\end{pmatrix} \overset{(i)'}{\rightarrow} \begin{pmatrix}\frac{1}{2}\\1\\\frac{1}{2}\end{pmatrix} = \mathbf{v}_2.$$

It is left as an exercise for the reader to perform similar calculations with the eigenvector (1)′. (First obtain an eigenvector $(1,2)'$ of B_1 then from this obtain the eigenvector $\mathbf{v}_3 = (2,2,1)'$ of A.)

1.5 Further Notes on the Power Method

The power method as described above can easily be programmed for a digital computer. This is particularly straightforward if the three essential stages of the process (power iteration, deflation and recovery of eigenvalues of the original matrix from those of the deflated matrices) are treated separately. Although the method has the merit of simplicity, it does have a number of important disadvantages which severely limit its practical application. The most important restriction is that the simple power iteration process will not converge when the matrix has as its eigenvalues of largest modulus a complex conjugate pair. (In this case, it is possible to find the eigenvalues by a method which considers the results of three consecutive iterations. See for example J. H. Wilkinson, *The Algebraic Eigenvalue Problem*.) Even when the matrix has real eigenvalues, the rate of convergence of the power method can be very slow although in some cases this can be enhanced by a suitable origin shift (see Exercise, 7, p. 23). Any program implementing the method must of course incorporate some reasonable limit (such as 100) to the number of iterations allowed together with an error indicator when the process fails to converge. In practice, even for a matrix with real eigenvalues well separated in modulus, the number of iterations needed by the power method can easily consume more computer time than more sophisticated methods such as the $Q-R$ algorithm.

References

1. WILKINSON, J. H. *The Algebraic Eigenvalue Problem*, O.U.P. (1965).

Exercises

1. If λ_i is an eigenvalue of a matrix A, $\mathbf{x}_i$ is said to be a *right eigenvector* of A if $A\mathbf{x}_i = \lambda_i \mathbf{x}_i$; $\mathbf{y}_i'$ is said to be a *left eigenvector* if $\mathbf{y}_i' A = \lambda_i \mathbf{y}_i'$ (or $A'\mathbf{y}_i = \lambda_i \mathbf{y}_i$). If

$$A = \begin{pmatrix}-4 & 4 & 2\\-3 & 2 & 3\\0 & 4 & -2\end{pmatrix},$$

find the eigenvalues and left and right eigenvectors of A. Show that, for $i \neq j$, $\mathbf{y}_i' \mathbf{x}_j = 0$.

2. If D is a diagonal matrix and U is an upper triangular matrix, show that in each case the eigenvalues are the diagonal elements. What are the forms of the corresponding eigenvectors of D and U respectively?

3. $\{\mathbf{x}_i, \mathbf{x}_2, \ldots, \mathbf{x}_n\}$ is a set of mutually orthogonal vectors ($\mathbf{x}_i'\mathbf{x}_j = 0$ for $i \neq j$). Show that for the symmetric matrix $A = \sum_{i=1}^{n} \alpha_i \mathbf{x}_i \mathbf{x}_i'$, the eigenvectors are $\mathbf{x}_1, \mathbf{x}_2, \ldots, \mathbf{x}_n$. What are the corresponding eigenvalues?

4. Find a symmetric matrix which has eigenvalues 1, 2, 4 and corresponding eigenvectors $(1, 0, -1)'$, $(1, 2, 1)'$ and $(1, -1, 1)'$.

5. The quadratic form $\phi(x_1, x_2, \ldots, x_n)$ is associated with the $n \times n$ symmetric matrix A. Assuming that A is positive definite if and only if all eigenvalues of A are positive, prove that $|A| > 0$. By considering $\phi(x_1, x_2, \ldots, x_k\ 0 \ldots, 0)$ prove that $|A_k| > 0$ for $k = 1, \ldots, n$, where A_k denotes the matrix which consists of the first k rows and columns of A. Prove also that all the diagonal elements of A are positive.

6. Use the power method to find, to two decimal places, the eigenvalue of largest modulus, and corresponding eigenvector of the matrix

$$A = \begin{pmatrix} -20 & -18 & -18 \\ 11 & 10 & 9 \\ -11 & -13 & -11 \end{pmatrix}.$$

7. Show that the application of the power method to the matrix

$$A = \begin{pmatrix} 5 & -10 & -10 \\ 3 & -1 & -3 \\ -3 & -4 & -2 \end{pmatrix}$$

produces, in the first few iterations, vectors in which the location of the element of largest modulus is constantly changing. Show also that if, after 4 or 5 iterations, an origin shift equal to the current estimate of the eigenvalue is made, the process starts to converge. (This matrix has two eigenvalues of equal modulus but opposite sign.)

8. Show that the matrix

$$A = \begin{pmatrix} -a_{n-1} & 1 & 0 & \cdots & & 0 \\ -a_{n-2} & 0 & 1 & 0 & \cdots & 0 \\ \vdots & \vdots & & \ddots & & \vdots \\ & & & & & 0 \\ -a_1 & & & & & 1 \\ -a_0 & 0 & \cdots & & 0 & 0 \end{pmatrix}$$

has characteristic polynomial $f(x) = x^n + a_{n-1}x^{n-1} + \ldots + a_1 x + a_0 = 0$. (This matrix is called the *companion matrix* of $f(x)$.)

9. Show that the matrices

$$A = \begin{pmatrix} 1 & -1 & 1 \\ 2 & -2 & 1 \\ 2 & -1 & 0 \end{pmatrix}, \qquad B = \begin{pmatrix} 2 & 1 & 2 \\ 3 & 4 & 10 \\ -2 & -2 & -5 \end{pmatrix}$$

have the same eigenvalues but are not similar matrices.

10. B is a real $n \times n$ matrix with n distinct real eigenvalues. Prove that if A is a matrix which commutes with B, then A has the same eigenvectors as B. What are the eigenvalues and eigenvectors of AB? Prove, conversely, that if B has the same eigenvectors as A then $AB = BA$.

11. By first finding the eigenvalues and eigenvectors of the coefficient matrix, solve the differential equations:

$$\frac{dx_1}{dt} = 3x_1 + 2x_2$$

$$\frac{dx_2}{dt} = -4x_1 + 3x_2 + 3x_3$$

$$\frac{dx_3}{dt} = 8x_1 - 2x_2 - 4x_3$$

CHAPTER 2

ERROR ANALYSIS

2.1 Introduction

This chapter is concerned with ideas and results which will be needed subsequently in considering the estimation of errors in the solution of sets of linear equations and in the evaluation of eigenvalues and eigenvectors. We deal first with rounding errors in basic arithmetic operations and with the concepts of perturbation, conditioning, and backward analysis. Secondly we look at errors in the formation of matrix products.

2.2 Computer Representation of Numbers

2.2.1

A number is held in a computer in binary form, as a string of 0's and 1's (a 'word'). Provided they are not too large, signed integers are held exactly; in ICL 1900 series FORTRAN an integer is held in a 24-bit word, which allows integers in the range

$$-838860 \leqslant I \leqslant 838859.$$

Moreover, integer arithmetic is carried out exactly, so long as it nowhere involves operations which may have a non-integral result and so long as the integers produced do not become too large ('overflow').

There are areas of computation which involve only integer arithmetic but generally the solution of a set of linear equations, or the extraction of an eigenvalue, will involve non-integral numbers. This may be the case even when the original data and the results produced are all integers. Accordingly, we use integers here only as labels, suffices, counters, etc.

Numbers not held as integers are referred to as *real numbers*. They may be held in either "fixed point form" or in "floating point form". For example, the fraction 1/16 may be written as a decimal either as 0.0625 (fixed point) or as 0.625×10^{-1} (floating point); the corresponding binary representations are 0.0001 (fixed point) and 0.1×2^{-3} (floating point).

In fixed point form the position of the binary or the decimal point in a word representing a real number is predetermined, and hence the range and *absolute* accuracy of representation is fixed. Almost invariably, numbers in fixed point working are taken to have no digits before the point, that is, to lie in the range $-1 \leqslant x < 1$. If a fixed point number is held in a word of length t bits (plus 1 bit for the sign), then it may be held with an error not exceeding

$$\tfrac{1}{2} \times 2^{-t} = 2^{-(t+1)} \text{ in modulus.}$$

A number like 2/3, as is well known, cannot be represented exactly either in decimal or in binary form. It is important to note, however, that there are exact decimal fractions which do not admit an exact representation in binary. For example, the exact decimal fraction 0.26 has the following, correctly rounded, representation as a ten bit binary number: 0.0100001011. This binary number is (exactly) equivalent to the decimal

$$0.259765625.$$

Hence, the binary representation of 0.26 has a rounding error (in decimal form) equal to 0.000234375.

The maximum rounding error which could occur is given by

$$2^{-11} = .00048828125.$$

Thus, ten binary digits will represent a number with an error a little less than 0.0005, and hence with an accuracy equivalent to three decimal places.

For a given word size, fixed point representation is capable of providing greater accuracy than is floating point. However, to take advantage of this, the calculation needs to be scaled. This is not only to ensure that no number occurs which is greater than one in modulus (to avoid overflow) but also that no number becomes too small. For example, if we work to six decimal places and a number becomes as small as (say) 10^{-4}, then the *relative* error in this number can be as large as $\tfrac{1}{2}(10^{-6}/10^{-4}) = \tfrac{1}{2}\%$. Relative errors of this size may invalidate a calculation completely. Clearly, it is not always easy to carry out this scaling, and it is simpler to accept the slightly smaller peak accuracy of floating-point representation.

2.2.2 Floating Point Numbers

Binary floating-point numbers are held in the form $x = a \times 2^b$; a is called the *mantissa*, and b the *exponent*.

Mantissa: In "standard floating-point" the mantissa is restricted to

$$\tfrac{1}{2} \leqslant |a| < 1.$$

If a is represented by a sign plus t bits then it will be held with an error not greater than $2^{-(t+1)}$. Since it is never less than $\tfrac{1}{2}$ in modulus, its relative error will be no more than $2^{-(t+1)}/\tfrac{1}{2} = 2^{-t}$.

We can express the computer representation of a in the form

$$\{a\} \equiv a(1 + \epsilon), \qquad \text{where } |\epsilon| \leqslant 2^{-t}.$$

In 1900 FORTRAN, we have $t = 37$, so that a is held with a relative accuracy of approximately eleven decimal digits ($2^{-t} = 2^{-37} \approx 7 \times 10^{-12}$).

Exponent: The exponent is an integer and is therefore held exactly, provided it is not too large. In 1900 FORTRAN, 9 bits are used to store $(256 + b)$, and so b lies in the range $-256 \leqslant b < 256$.

The allowable range of values of a real number x in 1900 FORTRAN is therefore: $2^{255} > |x| \geqslant \tfrac{1}{2} \times 2^{-256}$, which is approximately

$$2.9 \times 10^{75} > |x| \geqslant 8.6 \times 10^{-77}.$$

Provided $|x|$ is in this range there is no inaccuracy due to the exponent, and x itself is stored correct to about 11 significant figures:

$$\{x\} \equiv x(1 + \epsilon), \qquad \text{where } |\epsilon| \leqslant 2^{-t}. \tag{1}$$

If $|x|$ is less than 2^{-257} then x is stored as zero, and hence with an absolute error not greater than 2^{-257}. We may reasonably suppose that the calculation is scaled so that this absolute error is negligible.

2.3 Rounding Errors in Floating-point Arithmetic

2.3.1 Multiplication and Division

In floating-point arithmetic multiplication and division are comparatively simple operations. Thus, if

$$\{x_1\} = a_1 2^{b_1} \qquad \text{and} \qquad \{x_2\} = a_2 2^{b_2}$$

then we form $(a_1 a_2)2^{(b_1+b_2)}$ by multiplying together the mantissa and adding together the exponents. Since $\{x_1\}$ and $\{x_2\}$ are in standard form we have

$$\tfrac{1}{4} \leqslant |a_1 a_2| < 1,$$

and the adjustment required to put the product in standard form consists at most of moving the mantissa $a_1 a_2$ up one place and reducing the exponent $(b_1 + b_2)$ by 1. Almost certainly the only error that will arise is when $a_1 a_2$

(requiring $2t$ bits for its exact expression) is rounded to t bits. The fractional error introduced is thus not greater than 2^{-t} in modulus, and we can write

$$\{x_1x_2\} = \{x_1\}\{x_2\}(1+\epsilon), \qquad |\epsilon| \leqslant 2^{-t}. \tag{2}$$

Similar remarks apply to division:—

$$\{x_1/x_2\} = (\{x_1\}/\{x_2\})(1+\epsilon), \qquad |\epsilon| \leqslant 2^{-t}. \tag{3}$$

2.3.2 Addition and Subtraction

Note first that subtraction consists essentially of changing the sign and adding:

$$\text{SUBTRACTION of } x \equiv \text{ADDITION of } (-x).$$

Since change of sign should not involve any rounding error, it is enough to consider addition.

To form the sum of $\{x_1\} = a_1 2^{b_1}$ and $\{x_2\} = a_2 2^{b_2}$, we first need to arrange for the numbers to have a common exponent. Suppose that $b_1 \geqslant b_2$. Then:—

$$a_1 2^{b_1} + a_2 2^{b_2} = [a_1 + a_2 2^{-(b_1-b_2)}]\, 2^{b_1},$$

where $2^{-(b_1-b_2)} \leqslant 1$ and hence $a_2 2^{-(b_1-b_2)}$ is a proper fraction. Thus the mantissa of $\{x_2\}$ is moved down $(b_1 - b_2)$ places and added to a_1. If necessary, the result is then shifted, so that its modulus lies in the range $\frac{1}{2} \leqslant |a| < 1$, and b_1 is adjusted accordingly.

Usually the shift-add-shift operation is carried out exactly in a double length accumulator, only finally being followed by rounding to t bits. The resulting mantissa is correct to t bits and there is a final relative error of not more than 2^{-t}. We may write

$$\{x_1 + x_2\} = (\{x_1\} + \{x_2\})(1+\epsilon), \qquad |\epsilon| \leqslant 2^{-t} \tag{4}$$

If the computer does not use a double-length register for the formation of the mantissa of the sum then the relative error in the result may be very large. For example, let

$$\{x_1\} = (2^{-1} + 2^{-t})2^0 \qquad \text{and} \qquad \{x_2\} = -(1 - 2^{-t})2^{-1}.$$

With a double-length register the computer forms

$$a_1 + a_2 2^{-(b_1-b_2)} = (\tfrac{1}{2} + 2^{-t}) - (1 - 2^{-t})2^{-1} = 2^{-t} + 2^{-(t+1)}$$

and then moves it up $(t-1)$ places to give, in standard form, the exact result $\frac{3}{4} \times 2^{1-t}$. Without a double-length register, $a_2 2^{-(b_1-b_2)}$ must be rounded to t bits before being added to a_1. This gives a final result of 2^{-t}, which is in error by $33\frac{1}{3}\%$. Such a result appears at first to be quite unacceptable, but it is basically due to the cancellation taking place. Such cancellation will always have a similar effect, even if the computer addition is exact. To make this point clear, we need to undertake the more extensive analysis of the following sections.

2.4 Forward and Backward Analysis

2.4.1 Forward Analysis

We are concerned in the end not with the relative error produced by a single operation but with the relative error in a computer number when compared with the ideal, exact, result. That is to say, we wish to consider

$$\frac{\{x_1 + x_2\} - (x_1 + x_2)}{x_1 + x_2}.$$

Unless the computation is grossly in error it will make little difference whether we use $(x_1 + x_2)$ or $(\{x_1\} + \{x_2\})$ as a divisor. Replacing $(x_1 + x_2)$ by $(\{x_1\} + \{x_2\})$ in the first term, we may write

$$\frac{\{x_1 + x_2\} - (x_1 + x_2)}{x_1 + x_2} \approx \frac{\{x_1 + x_2\} - (\{x_1\} + \{x_2\})}{\{x_1\} + \{x_2\}} + \frac{\{x_1\} - x_1}{x_1 + x_2} + \frac{\{x_2\} - x_2}{x_1 + x_2} \quad (5)$$

The first term on the right-hand side is the fractional error produced by the computer's addition process and may be made small. However in the second and third terms errors arising from previous operations may be greatly magnified if $(x_1 + x_2)$ is small, and the less accurate machine addition does no more than add further errors of the same type.

In any case, it becomes impossible to pursue the bounds of relative error through a sequence of arithmetic operations which include additions (or subtractions) except in terms of the results at each stage, and this rapidly becomes prohibitively complicated. The way out of the difficulty is to introduce the idea of "backward analysis", which we proceed to develop in the next section.

2.4.2 Backward Analysis

Continuing with the analysis of the addition of a pair of numbers, we write in place of equation (4):

$$\{x_1 + x_2\} = \{x_1\}(1 + \epsilon_1) + \{x_2\}(1 + \epsilon_2) \quad (6)$$

where $|\epsilon_1|$ and $|\epsilon_2|$ are not greater than 2^{-t}. This equation can be expected to hold for any computer. If we suppose first that x_1 and x_2 have been read directly into the computer as data and that the addition considered is the first operation in which they are involved, then

$$\{x_1\} = (1 + \nu_1)x_1 \qquad \text{and} \qquad \{x_2\} = (1 + \nu_2)x_2$$

where $|\nu_1| \leqslant 2^{-t}$ and $|\nu_2| \leqslant 2^{-t}$. Equation (6) then becomes:

$$\begin{aligned}\{x_1 + x_2\} &= x_1(1+\nu_1)(1+\epsilon_1) + x_2(1+\nu_2)(1+\epsilon_2) \\ &\approx x_1(1+\nu_1+\epsilon_1) + x_2(1+\nu_2+\epsilon_2),\end{aligned} \tag{7}$$

neglecting very small terms of order 2^{-2t}. We can interpret this last equation to mean that the computer number $\{x_1 + x_2\}$ is the result of the *exact* addition, not of x_1 and x_2, but of $x_1(1+\delta_1)$ and $x_2(1+\delta_2)$, where $\delta_1 = \nu_1 + \epsilon_1$ and $\delta_2 = \nu_2 + \epsilon_2$. We thus equate the effect of the errors due to storing the numbers and then adding them with the effect of introducing initial "perturbations" into the data, these perturbations being relative errors with bounds given in this case by

$$|\delta_1| \leqslant |\epsilon_1| + |\nu_1| \leqslant 2.2^{-t} \qquad \text{and} \qquad |\delta_2| \leqslant |\epsilon_2| + |\nu_2| \leqslant 2.2^{-t}.$$

We can usually make such a correspondence between rounding errors and equivalent initial perturbations, and estimate bounds for these perturbations. The process is known as "Backward Analysis", and it reduces the estimation of the effect of rounding error to that of the effects of perturbations in the inputs to the calculations. But very often this second estimation has to be made anyway – certainly for those input quantities which are obtained by experiment. In fact, it may be immediately obvious that rounding errors are unimportant compared with experimental errors. Again, the effects of small changes in input quantities may be obvious from physical considerations, and backward analysis will enable this physical insight to assist in the problem of analysing the effect of rounding errors. But even if there are no experimental errors or physical changes equivalent to the rounding-error perturbations, it is usually possible to estimate the effect of these perturbations, and to decide whether or not these effects will be serious. A calculation in which small relative changes in the input quantities produce large relative changes in the output is said to be "ill-conditioned", and a number which measures the degree of ill-conditioning is called a "Condition Number".

2.4.3 Example

We illustrate the idea of conditioning by considering once more the addition of two numbers. We have obtained the result

$$\{x_1 + x_2\} = x_1(1+\delta_1) + x_2(1+\delta_2)$$

with a relative error given by

$$|\delta| = \left| \left(\frac{x_1}{x_1+x_2}\right)\delta_1 + \left(\frac{x_2}{x_1+x_2}\right)\delta_2 \right| \leqslant \frac{|x_1|}{|x_1+x_2|}\,|\delta_1| + \frac{|x_2|}{|x_1+x_2|}\,|\delta_2|.$$

If $\dfrac{|x_1|}{|x_1+x_2|}$, or $\dfrac{|x_2|}{|x_1+x_2|}$, (or both) is large compared with unity the problem is ill-conditioned; if they are both of the order of unity (or less) the problem is well-conditioned. Either of the numbers

$$\frac{|x_1| + |x_2|}{|x_1 + x_2|} \quad \text{or} \quad \frac{\max(|x_1|, |x_2|)}{|x_1 + x_2|}$$

could be used as a condition number. Note that the conditioning and the condition number are alike independent of the magnitudes of δ_1 and δ_2, and of whether these are original errors in x_1 and x_2, or equivalent perturbations replacing errors.

Clearly multiplication and division may be brought into the framework of backward analysis. Equations (2) and (3) may be interpreted as showing that the rounding error is equivalent to introducing a perturbation in one of the input quantities. If, as before

$$\{x_1\} = (1 + \nu_1)x_1 \quad \text{and} \quad \{x_2\} = (1 + \nu_2)x_2\text{, then}$$
$$\{x_1x_2\} = x_1(1 + \nu_1)x_2(1 + \nu_2)(1 + \epsilon) \approx [x_1(1 + \nu_1)]\,[x_2(1 + \nu_2 + \epsilon)].$$

In general, if

$$\{x_1x_2\} = x_1(1 + \delta_1).x_2(1 + \delta_2),$$

the relative error is given by

$$|\delta| = \left|\frac{\{x_1x_2\} - x_1x_2}{x_1x_2}\right| = |\delta_1 + \delta_2 + \delta_1\delta_2|$$
$$\approx |\delta_1 + \delta_2| \leqslant |\delta_1| + |\delta_2|.$$

There is never any magnification of accumulated errors, and multiplication *by itself* is always well-conditioned. A similar result holds for division.

2.5 Extended Arithmetic

2.5.1 Multiplication

The derivation of bounds for the rounding errors arising from a string of multiplications is straightforward. We have

$$\{x_1x_2\} = \{x_1\}\{x_2\}(1 + \epsilon_1)$$
$$\{x_1x_2x_3\} = \{x_1\}\{x_2\}(1 + \epsilon_1)\{x_3\}(1 + \epsilon_2),$$

and so on, so that

$$\{x_1x_2 \ldots x_n\} = \{x_1\}\{x_2\} \ldots \{x_n\}(1 + E),$$

where

$$(1 - 2^{-t})^{n-1} \leqslant 1 + E \leqslant (1 + 2^{-t})^{n-1}$$

Clearly, division is similar, and in general for mixed multiplication and division we have

$$\left\{\frac{x_1x_2\ldots x_n}{y_1y_2\ldots y_m}\right\}=\frac{\{x_1\}\{x_2\}\ldots\{x_n\}}{\{y_1\}\{y_2\}\ldots\{y_m\}}\,(1+E) \tag{8}$$

where $(1-2^{-t})^{m+n-1}\leqslant 1+E\leqslant(1+2^{-t})^{m+n-1}$. Assuming that $(m+n)$ is small compared with 2^t, ($2^{30}\approx 10^{10}$, and so this is likely), we can write, with negligible error,

$$|E|\leqslant(m+n-1)2^{-t},$$

the error bounds being proportional to the number of operations carried out.

2.5.2 Addition and Subtraction

The analysis of the errors in the sum of a number of terms when accurate (double-length register) addition is used is as follows:

$$\{x_1+x_2\}=(\{x_1\}+\{x_2\})(1+\epsilon_2),$$
$$\begin{aligned}\{x_1+x_2+x_3\}&=(\{x_1\}+\{x_2\})(1+\epsilon_2)+\{x_3\}(1+\epsilon_3)\\&=\{x_1\}(1+\epsilon_2)(1+\epsilon_3)+\{x_2\}(1+\epsilon_2)(1+\epsilon_3)+\{x_3\}(1+\epsilon_3),\end{aligned}$$

and, finally,

$$\{x_1+x_2+\ldots+x_n\}=(1+\eta_1)\{x_1\}+(1+\eta_2)\{x_2\}+\ldots+(1+\eta_n)\{x_n\},$$

where

$$1+\eta_1=(1+\epsilon_2)(1+\epsilon_3)\ldots(1+\epsilon_n),$$

and

$$1+\eta_r=(1+\epsilon_r)(1+\epsilon_{r+1})\ldots(1+\epsilon_n)\quad\text{for}\quad r>1.$$

The only difference if the less accurate addition is used is that no longer do the same values of ϵ occur in more than one η_r; the numbers of factors in the expressions are the same, as are the bounds for each of the ϵ. Hence, in either case,

$$\begin{aligned}&(1-2^{-t})^{n-1}\leqslant(1+\eta_1)\leqslant(1+2^{-t})^{n-1},\text{ and}\\&(1-2^{-t})^{n-r+1}\leqslant(1+\eta_r)\leqslant(1+2^{-t})^{n-r+1},\quad r>1.\end{aligned} \tag{9}$$

With negligible error we may use equation (9) for $r=1$ as well as for $r>1$. If, in addition, n is small compared with 2^t, then

$$|\eta_r|\leqslant(n-r+1)2^{-t}. \tag{10}$$

A bound for the error is given by

$$|\{\Sigma x_r\}-\Sigma\{x_r\}|=|\Sigma\eta_r\{x_r\}|\leqslant\Sigma|\eta_r|\,|x_r|. \tag{11}$$

We notice that the bounds for $|\eta_r|$ decreases as r increases, and hence we may keep the bounds in (11) as small as possible by arranging $x_1, x_2 \ldots x_n$ in increasing order of absolute magnitude. That is, small terms should be added first, large terms being introduced later.

2.5.3 Example

Using floating-point decimal arithmetic with a four-digit mantissa, add together the following numbers in the order given:

$$0.4605 + 0.4600 + 0.4105 + 1.049 + 26.04 + 62.14 + 911.3$$

After each addition, the current result is rounded to four significant digits, so that we successively form,

$$\begin{aligned}
0.4605 + 0.4600 &= 0.9205\\
0.9205 + 0.4105 &= 1.341\\
1.341 + 1.049 &= 2.390\\
2.390 + 26.04 &= 28.43\\
28.43 + 62.14 &= 90.57\\
90.57 + 911.3 &= 1001.87 \approx 1002.
\end{aligned}$$

Taking the terms in the reverse order we have,

$$\begin{aligned}
911.3 + 62.14 &= 973.44 &&\approx 973.4\\
973.4 + 26.04 &= 999.44 &&\approx 999.4\\
999.4 + 1.049 &= 1000.449 &&\approx 1000\\
1000 + 0.4105 &= 1000.4105 &&\approx 1000\\
1000 + 0.4600 &= 1000.4600 &&\approx 1000\\
1000 + 0.4605 &= 1000.4605 &&\approx 1000.
\end{aligned}$$

The first result is correct to four figures. The second is 2 out in the fourth place and is correct only to three figures. The effect can be very much more marked if a large number of terms is involved.

2.5.4 Double-length Arithmetic

It is sometimes useful to carry out a particularly sensitive part of a calculation in doube-length arithmetic; that is, using a mantissa of $2t$ bits (or approximately $2t$ bits) instead of one of t bits. Unless the double-length calculations are particularly long and ill-conditioned the only error produced will be that due to the final step of rounding to t bits before proceeding with the rest of the calculation. In the context of matrix calculations it is frequently necessary to compute expressions such as $\sum_{r=1}^{n} x_r y_r$, and it is often useful to use double-length arithmetic for this. We are thus able to write

$$\left\{\sum_{r=1}^{n} x_r y_r\right\} = (1+\epsilon) \sum_{r=1}^{n} \{x_r\}\{y_r\}.$$

For the purpose of backward analysis we can then associate the $(1 + \epsilon)$ either with each of the $\{x_r\}$, or with each of the $\{y_r\}$.

2.6 Statistical Error Bounds

Consider again the multiplication of a number of factors:

$$\{x_1 x_2 \ldots x_n\} = \{x_1\}\{x_2\} \ldots \{x_n\}(1+E)$$

where $|E| \leqslant (n-1)2^{-t}$. This bound is an absolute limit, and it is very unlikely to be attained. We have, approximately,

$$E = \sum_{i=2}^{n} \epsilon_i \qquad \text{where } |\epsilon_i| \leqslant 2^{-t}.$$

If we assume that the ϵ_i are statistically independent and are each distributed uniformly in the range $-2^{-t} \leqslant \epsilon_i \leqslant 2^{-t}$, then for fairly large n (say $n > 10$), E is approximately normally distributed with standard deviation

$$\sigma = \sqrt{n-1} \times \text{s.d. of } \epsilon_i = \sqrt{\frac{n-1}{3}}\, 2^{-t}.$$

With about 99.7% probability, E deviates from zero by less than three standard deviations. That is, almost certainly we have

$$|E| \leqslant \sqrt{3(n-1)}\, 2^{-t}.$$

Comparing this with the previous absolute error bound, it is as if the basic rounding error is $\sqrt{\frac{3}{n-1}}\, 2^{-t}$ instead of 2^{-t}, where $(n-1)$ is the number of basic operations (in this case, multiplications) involved in the computation. Similar results hold for other arithmetic operations, and whilst we shall continue to use absolute bounds, we can bear in mind that in some situations at least we could fairly safely reduce these.

2.7 Matrix Calculation

2.7.1 Introduction

We shall presently illustrate the previous analysis by considering errors in the formation of the product of two matrices. We have indicated that we shall use backward analysis to estimate the effect of rounding errors, reducing the analysis to the problem of estimating the effect of perturbations in one or both of the original matrices. We shall see by example that the problem may be ill-conditioned.

Consider the ideal calculation:

$$\begin{pmatrix} 1 & 2 & 3 \\ 3 & 1 & 2 \\ -1 & 3 & 4 \end{pmatrix} \begin{pmatrix} 2 \\ 6 \\ -5 \end{pmatrix} = \begin{pmatrix} -1 \\ 2 \\ -4 \end{pmatrix} \tag{12}$$

Suppose that the elements of the first matrix on the left-hand side are affected by perturbations of the order of a few per cent, and form the product using the perturbed matrix. As a specific example, consider the following:–

$$\begin{pmatrix} 1.02 & 2.04 & 2.96 \\ 3.03 & 1.05 & 1.95 \\ -1.02 & 2.95 & 4.05 \end{pmatrix} \begin{pmatrix} 2 \\ 6 \\ -5 \end{pmatrix} = \begin{pmatrix} -0.52 \\ 2.61 \\ -4.59 \end{pmatrix}.$$

The percentage errors in the first matrix and in the product matrix are shown below.

$$\begin{pmatrix} 2\% & 2\% & 1.3\% \\ 1\% & 5\% & 2.5\% \\ 2\% & 1.7\% & 1.25\% \end{pmatrix} \begin{pmatrix} 48\% \\ 30.5\% \\ 15\% \end{pmatrix}$$

Clearly, there is considerable magnification in the relative errors in this case. On the other hand, a different set of perturbations, of similar magnitudes, may have a quite different effect:

$$\begin{pmatrix} 1.02 & 2.06 & 3.08 \\ 3.02 & 1.01 & 2.02 \\ -1.02 & 3.04 & 4.04 \end{pmatrix} \begin{pmatrix} 2 \\ 6 \\ -5 \end{pmatrix} = \begin{pmatrix} -1.00 \\ 2.00 \\ -4.00 \end{pmatrix}$$

In this case, the perturbations leave the original, exact, result unaltered.

Before proceeding to the general case, it is worth examining this simple example in more detail. It is sufficient initially to use only one row of the (3 x 3) matrix. We may take, for example, the product

$$(1, 2, 3)\,(2, 6, -5)' = -1. \tag{13}$$

Each of the vectors on the left-hand side of (13) may be interpreted as representing a "physical" vector in three-dimensional space, and the matrix product of the vector matrices is then the scalar product, in the usual, geometric, sense, of the physical vectors. This is it s turn is equal to the product of the lengths of the vectors and the cosine of the angle between them. It will be small if the vectors are nearly mutually perpendicular, even if the vectors themselves are not small. In the above example, the vectors are of lengths 3.74 and 8.06 respectively, and they have an angle of about 92° between them. The product of the vector lengths is fairly large (30.17), but cos 92° is −0.0349, and multiplying by this gives (approximately) −1.0, which is comparatively small. The multiplication of two vectors which are nearly perpendicular is ill-conditioned in the same way as is the addition of two numbers nearly equal in magnitude but of opposite sign. In equation (12) all three rows of the first matrix were chosen to be nearly perpendicular to the column vector $(2\ \ 6\ \ -5)'$ and, as a result, the calculation of the product is ill-conditioned. To illustrate this, perturbations were chosen to be nearly parallel to $(2\ \ 6\ \ -5)'$ and hence to yield comparatively large contributions to the product in spite of their small magnitudes.

As a condition number for the product $(\mathbf{x}'\mathbf{y})$ of two vectors, we could use any function of the angle, θ, between $\mathbf{x}$ and $\mathbf{y}$ which becomes large as θ approaches 90°. The most obvious and convenient is

$$K = \frac{1}{|\cos\theta|} = \frac{\|\mathbf{x}\|\ \|\mathbf{y}\|}{\|\mathbf{x}'\mathbf{y}\|} \tag{14}$$

where $\|\mathbf{x}\| = +\sqrt{x_1^2 + x_2^2 + x_3^2}$, the length of the vector (x_1, x_2, x_3) in the usual sense of the term. An expression of the same form applies even when more than three dimensions are involved, although there is no longer any direct analogue of the angle θ. For matrix products in general, however, some generalisation of the idea of length, or magnitude, must be defined.

2.7.2 Matrix Norms

Any numerical function of the elements of a square matrix $A = [a_{ij}]$ is called a *matrix norm* if it satisfies the following properties:

$$\begin{aligned} &\|A\| > 0 \quad \text{unless } A = 0,\\ &\|kA\| = |k|\ \|A\|, \text{ where } k \text{ is any scalar,}\\ &\|A + B\| \leqslant \|A\| + \|B\|\\ &\|AB\| \leqslant \|A\|\ \|B\|. \end{aligned} \tag{15}$$

Such a function will serve, in some sense, as a more or less convenient measure of the "size" or "magnitude" of the matrix. A fuller discussion of the concept of norm is given in the Appendix. For the moment, we shall simply list some of the more important norms which may be defined:

(i) *Euclidean norm.* $$\|A\|_E = \sqrt{\sum_{i=1}^{n}\sum_{j=1}^{n} |a_{ij}|^2} \tag{16}$$

(ii) ℓ_2 *norm.* $$\|A\|_2 = \sqrt{\{\text{the largest eigenvalue of } \bar{A}'A\}} \tag{17}$$

(iii) ℓ_1 *norm.* $$\|A\|_1 = \max_j \sum_i |a_{ij}| \tag{18}$$

(iv) ℓ_∞ *norm.* $$\|A\|_\infty = \max_i \sum_j |a_{ij}|. \tag{19}$$

The Euclidean norm is perhaps the one which most readily appears in the light of a generalisation of the most familiar norm used for vectors:

$$\|\mathbf{x}\| = \sqrt{\sum_{i=1}^{n} |x_i|^2} \tag{20}$$

However, it turns out to be of less theoretical importance than the other norms listed above, particularly the ℓ_2 norm. Further, note that for the unit matrix I we have $\|I\|_E = \sqrt{n}$, where n is the dimension of I, while for each of the other norms listed $\|I\| = 1$.

2.7.3 Perturbations in Matrix Products

We suppose that we wish to form the product

$$C = AB,$$

but that we only have available the perturbed matrices $A + \delta A$ and $B + \delta B$, from which we form

$$C + \delta C = (A + \delta A)(B + \delta B).$$

The error in C due to the perturbations in A and B is

$$\delta C = A.\delta B + \delta A.B + \delta A.\delta B.$$

We shall try to find a bound for the matrix δC. Using the third and fourth of equations (15), we find

$$\|\delta C\| \leqslant \|A\|.\|\delta B\| + \|\delta A\|.\|B\| + \|\delta A\|.\|\delta B\|. \tag{21}$$

We are prepared to tolerate moderate values of $\|\delta C\|$ if $\|C\|$ itself is large, but we shall certainly require $\|\delta C\|$ to be very small if $\|C\|$ is small. We consider, therefore, a sort of relative error given by the ratio of $\|\delta C\|$ to $\|C\|$, and write this in terms of the relative perturbations $\|\delta A\|/\|A\|$ and $\|\delta B\|/\|B\|$ by dividing equation (21) by $\|C\| = \|AB\|$ to obtain

$$\frac{\|\delta C\|}{\|C\|} \leqslant \frac{\|A\|.\|B\|}{\|AB\|}\left[\frac{\|\delta B\|}{\|B\|} + \frac{\|\delta A\|}{\|A\|} + \frac{\|\delta A\|}{\|A\|}.\frac{\|\delta B\|}{\|B\|}\right].$$

We expect the relative perturbations to be small, so that we can ignore the last term on the right-hand side, and write finally

$$\frac{\|\delta C\|}{\|C\|} \leqslant \frac{\|A\|.\|B\|}{\|AB\|}\left[\frac{\|\delta A\|}{\|A\|} + \frac{\|\delta B\|}{\|B\|}\right]. \tag{22}$$

If the quantity

$$K = \frac{\|A\|.\|B\|}{\|AB\|} \tag{23}$$

is large, the relative perturbations may be seriously magnified, and we can therefore use K as a condition number. (Note that if A and B are replaced by row and column matrices respectively, expression (23) is identical with that previously given in equation (14).)

Since
$$\|AB\| \leqslant \|A\| \, . \, \|B\|$$

for any two matrices, it follows that K is never less than unity. We should note that this does not mean that the relative error in the product is never less than the sum of the relative perturbations. However large K may be, there will be some perturbations in A and B which produce only very small changes in C. On the other hand, there will be other perturbations which make the two sides of equation (22) approximately equal.

Example

Let $A = \begin{pmatrix} 1 & -1 & 1 & -1 \\ 1 & -1 & -1 & 1 \\ -1 & 1 & 1 & 1 \\ -0.9 & 0.9 & 0.9 & -0.9 \end{pmatrix}$ and $B = \begin{pmatrix} 1 & 1 & -1 & -1 \\ 1 & 1 & -1 & -1 \\ 9 & -1 & 1 & -1 \\ 0.9 & -0.9 & 0.9 & -0.9 \end{pmatrix}$

Then
$$C = AB = \begin{pmatrix} 0.1 & -0.1 & 0.1 & -0.1 \\ -0.1 & 0.1 & -0.1 & 0.1 \\ -0.1 & 0.1 & -0.1 & 0.1 \\ 0.09 & 0.09 & 0.09 & -0.09 \end{pmatrix}$$

For each of the four different norms introduced, the following table shows the norms of A, B and C and the corresponding condition numbers.

Norm	A	B	C	K
ℓ_1	3.9	3.9	0.39	39
ℓ_∞	4.0	4.0	0.40	40
Euclidean	3.9	3.9	0.39	39
ℓ_2	2.8	2.8	0.28	28

The condition number varies somewhat with the particular norm used, but whether we take it as 29 or 40, it suggests that small errors in A or B may produce comparatively large errors in C. To take a specific example, let

$$\delta A = 10^{-2} \begin{pmatrix} 1 & 1 & 1 & 1 \\ 1 & 1 & 1 & 1 \\ 1 & 1 & 1 & 1 \\ 1 & 1 & 1 & 1 \end{pmatrix}, \quad \delta B = 0.$$

Then
$$(A + \delta A)B = \begin{pmatrix} 0.139 & -0.099 & 0.099 & -0.139 \\ -0.061 & 0.101 & -0.101 & 0.061 \\ -0.061 & 0.101 & -0.101 & 0.061 \\ 0.129 & -0.089 & 0.089 & -0.129 \end{pmatrix}$$

The error in the product is thus

$$\delta C = \begin{pmatrix} 0.039 & 0.001 & -0.001 & -0.039 \\ 0.039 & 0.001 & -0.001 & -0.039 \\ 0.039 & 0.001 & -0.001 & -0.039 \\ 0.039 & 0.001 & -0.001 & -0.039 \end{pmatrix}$$

The relative errors in the elements of A are all about 1%, whilst the relative errors in some elements of C are about 40%. The sort of magnification allowed by the condition number does in this case take place. In terms of the matrix norms, the following table shows the magnification actually occurring compared with the condition number.

Norm	$\|\delta A\|$	$\|\delta C\|$	$\dfrac{\|\delta C\|}{\|C\|} \Big/ \dfrac{\|\delta A\|}{\|A\|}$	K
ℓ_1	0.04	0.156	39	39
ℓ_∞	0.04	0.080	20	40
Euclidean	0.04	0.100	25	39
ℓ_2	0.04	0.100	25	28

The reader may confirm that in spite of the fairly large condition number, the perturbation:

$$\delta A = 10^{-2} \begin{pmatrix} 1 & -1 & 1 & -1.1 \\ 1 & -1 & -1 & 1.1 \\ -1 & 1 & -1 & 1.1 \\ -1 & 1 & -1 & -1.1 \end{pmatrix}, \quad \delta B = 0$$

has very little effect on the value of the product.

2.7.4 Rounding Errors in Matrix Products

We shall try to replace the rounding errors which may occur in forming a matrix product by equivalent perturbations in the two matrices to be multiplied together. If double-length accumulation of sums of products is available, there will be very little rounding and we assume that double-length arithmetic is not used. Let A be an $(\ell \times m)$ matrix and B an $(m \times n)$ matrix. For $i = 1, 2 \ldots \ell$ and

$j = 1, 2 \dots n$: the elements of the product C are given by

$$c_{ij} = \sum_{k=1}^{m} a_{ik} b_{kj}.$$

From Section 2.4, the computed element is given by

$$\{c_{ij}\} = \sum_{k=1}^{m} (1 + \zeta_k) a_{ik} b_{kj}, \qquad |\zeta_k| \leqslant (m - k + 2) 2^{-t}.$$

This result is equivalent to that obtained by computing exactly with an exact matrix B, but using a perturbed matrix $A + \delta A$, where the elements of δA satisfy the inequality

$$|\delta a_{ij}| \leqslant (m - j + 2) 2^{-t} \, |a_{ij}|$$

Now, in Section 2.7.2 the concept of matrix norm was introduced specifically for *square* matrices. However, a glance at the definitions of the particular norms given in equations (16), (17), (18),and (19) shows that they generalise quite readily to apply to rectangular matrices. Accordingly, it makes sense to try to convert the above inequality into an upper bound for the relative perturbation $\|\delta A\| / \|A\|$.

We consider each of the four norms in turn.

(i)
$$\|\delta A\|_1 = \max_j \sum_i |\delta a_{ij}| \leqslant \max_j \sum_i (m - j + 2) 2^{-t} \, |a_{ij}|$$

$$= 2^{-t} \max_j (m - j + 2) \sum_i |a_{ij}|.$$

But
$$\max_j \sum_i |a_{ij}| = \|A\|_1$$

and
$$\max_j (m - j + 2) = (m + 1),$$

and hence
$$\max_j (m - j + 2) \sum_i |a_{ij}| \leqslant (m + 1) \|A\|_1$$

Therefore,
$$\frac{\|\delta A\|_1}{\|A\|_1} \leqslant (m + 1) 2^{-t}.$$

(ii) In exactly the same way:—

$$\frac{\|\delta A\|_\infty}{\|A\|_\infty} \leqslant (m + 1) 2^{-t}.$$

(iii)
$$\|\delta A\|_E = \sqrt{\left[\sum_i \sum_j (\delta a_{ij})^2\right]}$$

$$\leqslant 2^{-t} \sqrt{\left[\sum_i \sum_j (m - j + 2)^2 a_{ij}^2\right]}$$

$$\leqslant 2^{-t}(m+1)\sqrt{\left[\sum_i \sum_j a_{ij}^2\right]} = 2^{-t}(m+1)\|A\|_E.$$

(iv) There is no very useful bound for the ℓ_2 norm.

Assuming that we use something other than the ℓ_2 norm, we take $\|\delta A\|/\|A\| \leqslant (m+1)2^{-t}$ in equation (20) and obtain

$$\frac{\|\delta C\|}{\|C\|} \leqslant \frac{\|A\| \,.\, \|B\|}{\|AB\|}(m+1)2^{-t}. \tag{22}$$

For the example used in the last section; $m = 4$ and $\dfrac{\|A\| \,.\, \|B\|}{\|AB\|} = 40$ and we have $\dfrac{\|\delta C\|}{\|C\|} \leqslant 200 \times 2^{-t}$.

To guarantee a relative error of not more than, say, 1% we require

$$200 \times 2^{-t} \leqslant 0.01,$$

and hence we have to use about 15 bits for the floating-point arithmetic.

CHAPTER 3

THE SOLUTION OF LINEAR EQUATIONS BY ELIMINATION AND DECOMPOSITION METHODS

3.1 Introduction

Practical methods for the solution of systems of linear equations fall into two main classes. This chapter is concerned with *direct* methods, for which, in principle, a single application of a manipulative process suffices to give an exact solution. *Indirect* methods generally make repeated use of a rather simpler type of process to obtain successively improved approximations to the solution. Some important members of this family are examined in the next chapter, where it is pointed out that indirect methods are best reserved for use under rather special circumstances arising, typically, in the numerical solution of certain problems involving differential equations. For general-purpose use, the direct methods of this chapter are much more reliable.

Even though a direct method is designed to produce an exact solution, the limitations of computers make this an unattainable goal in practice, because of the occurrence of rounding errors during the computation. However, measures can be taken to ensure that these rounding errors have the least possible effect on the final answer, and we pay due attention to this matter in what follows.

3.2 Elimination Methods

3.2.1 Gauss's Method

All the methods of this chapter are based ultimately on the process of elimination of variables. The method of Gauss is a systematic elimination procedure whose object is the transformation of an initial rectangular system of equations into a triangular system which is much more easily solved.

Consider, for example, the following system of 4 equations in 4 variables:

$$\begin{aligned} 3x_1 + x_2 - 2x_3 - x_4 &= 3 \\ 2x_1 - 2x_2 + 2x_3 + 3x_4 &= -8 \\ x_1 + 5x_2 - 4x_3 - x_4 &= 3 \\ 3x_1 + x_2 + 2x_3 + 3x_4 &= -1. \end{aligned} \tag{1}$$

It is convenient to suppress all unnecessary symbols and to operate upon an array composed entirely of numerical elements:

$$\left(\begin{array}{cccc:c} 3 & 1 & -2 & -1 & 3 \\ 2 & -2 & 2 & 3 & -8 \\ 1 & 5 & -4 & -1 & 3 \\ 3 & 1 & 2 & 3 & -1 \end{array}\right).$$

Such an array is called an *augmented matrix.* It is only necessary to bear in mind that the columns represent, from left to right, coefficients of x_1, x_2, x_3 and x_4, and constants, while each row represents one of the original equations. The process of triangularisation makes use of the following *elementary row operations:*

(i) interchange of any two rows in the matrix,
(ii) multiplication of any row by a non-zero constant, and
(iii) addition of any two rows; with (ii), this permits addition of any non-zero multiple of one row to another.

Viewed as operations performed on the original system of equations, these procedures clearly do not affect the solution of that system. Their application to our augmented matrix enables us to generate other augmented matrices which are equivalent to the original in that the systems of equations they represent all have the same solution. The sequence of equivalent matrices generated by the Gauss process is this:

$$\left(\begin{array}{cccc:c} 3 & 1 & -2 & -1 & 3 \\ 2 & -2 & 2 & 3 & -8 \\ 1 & 5 & -4 & -1 & 3 \\ 3 & 1 & 2 & 3 & -1 \end{array}\right), \quad \begin{array}{l} R_1 \\ R_2 - \frac{2}{3}R_1 \\ R_3 - \frac{1}{3}R_1 \\ R_4 - R_1 \end{array} \quad \left(\begin{array}{cccc:c} 3 & 1 & -2 & -1 & 3 \\ 0 & -\frac{8}{3} & \frac{10}{3} & \frac{11}{3} & -10 \\ 0 & \frac{14}{3} & -\frac{10}{3} & -\frac{2}{3} & 2 \\ 0 & 0 & 4 & 4 & -4 \end{array}\right),$$

$$\begin{array}{l} R_1 \\ R_2 \\ R_3+\frac{7}{4}R_2 \\ R_4 \end{array} \begin{pmatrix} 3 & 1 & -2 & -1 & \vdots & 3 \\ 0 & -\frac{8}{3} & \frac{10}{3} & \frac{11}{3} & \vdots & -10 \\ 0 & 0 & \frac{5}{2} & \frac{23}{4} & \vdots & -\frac{31}{2} \\ 0 & 0 & 4 & 4 & \vdots & -4 \end{pmatrix}, \quad \begin{array}{l} R_1 \\ R_2 \\ R_3 \\ R_4-\frac{8}{5}R_3 \end{array} \begin{pmatrix} 3 & 1 & -2 & -1 & \vdots & 3 \\ 0 & -\frac{8}{3} & \frac{10}{3} & \frac{11}{3} & \vdots & -10 \\ 0 & 0 & \frac{5}{2} & \frac{23}{4} & \vdots & -\frac{31}{2} \\ 0 & 0 & 0 & -\frac{26}{5} & \vdots & \frac{104}{5} \end{pmatrix}.$$

Here the R's are row numbers, and the operations indicated are those which have been performed at each stage on the previous matrix. First, appropriate multiples of R_1 are subtracted from R_2, R_3 and R_4 so as to generate zeros in the C_1 (column 1) positions in those rows. This is clearly tantamount to elimination of x_1 from the last three equations. Next, multiples of R_2 in the modified (or *reduced*) matrix are subtracted from R_3 and R_4 to generate zeros in the C_2 positions of those rows, with the effect of eliminating x_2 from the last two equations. The process proceeds similarly, terminating when the last row contains only one non-zero element. The augmented matrix at this stage, in which the left-hand part is triangular, is equivalent to the system of equations

$$\begin{aligned} 3x_1 + x_2 - 2x_3 - x_4 &= 3 \\ -\frac{8}{3}x_2 + \frac{10}{3}x_3 + \frac{11}{3}x_4 &= -10 \\ \frac{5}{2}x_3 + \frac{23}{4}x_4 &= -\frac{31}{2} \\ -\frac{26}{5}x_4 &= \frac{104}{5}. \end{aligned}$$

This system is easily solved by *back-substitution*; the value of x_4 is obtained immediately, that of x_3 then follows from the third equation, x_2 from the second, and so on. The results are

$$x_4 = -4, \quad x_3 = 3, \quad x_2 = 2, \quad x_1 = 1.$$

It is important to realise that a computer could not reproduce the exact arithmetic used in the foregoing example; even such comparatively simple decimal fractions as $\frac{8}{3}$ and $\frac{26}{5}$ cannot be represented exactly as terminating binary fractions. The arithmetic of the computer, and therefore also the final solution, would be subject to rounding errors.

A convenient by-product of the Gauss process is the value of the determinant of the coefficient matrix of the original system of equations. Since the reduction to triangular form was effected entirely by subtraction of multiples of one row from another, the determinant of the left-hand part of the augmented matrix remains unchanged. The determinant of the final, triangular, reduced matrix is the product of its diagonal elements,

$$\Delta = 3 \times -\frac{8}{3} \times \frac{5}{2} \times -\frac{26}{5} = 104,$$

and this, then, is also the determinant of the original matrix. In general, the triangularisation method requires roughly $\frac{1}{3}n^3$ multiplications to calculate the determinant of an $n \times n$ matrix, whereas a direct brute-force expansion would require $n!$ multiplications. For large n, the computational advantage of the former method is clearly immense.

As we have seen, the basis of the Gauss method is the use at each stage of a particular element, generally termed the *pivot*, to generate zeros in its column in all the rows beneath it. In the process illustrated, the first pivot occupies the a_{11} position in the original matrix, the second pivot the a_{22} position of the first reduced matrix, and so on; all the pivots lie on the diagonal of the final reduced matrix. The chief failing of the basic Gauss method is that it is possible for zero pivots to arise, and that the process then terminates prematurely. For instance, if our original set of equations is written in a different order, we obtain

$$\begin{pmatrix} 3 & 1 & -2 & -1 & \vdots & 3 \\ 3 & 1 & 2 & 3 & \vdots & -1 \\ 2 & -2 & 2 & 3 & \vdots & -8 \\ 1 & 5 & -4 & -1 & \vdots & 3 \end{pmatrix}, \quad \begin{matrix} R_1 \\ R_2 - R_1 \\ R_3 - \frac{2}{3}R_1 \\ R_4 - \frac{1}{3}R_1 \end{matrix} \begin{pmatrix} 3 & 1 & -2 & -1 & \vdots & 3 \\ 0 & 0 & 4 & 4 & \vdots & -4 \\ 0 & -\frac{8}{3} & \frac{10}{3} & \frac{11}{3} & \vdots & -10 \\ 0 & \frac{14}{3} & -\frac{10}{3} & -\frac{2}{3} & \vdots & 2 \end{pmatrix},$$

in which the second pivot has proved to be zero. In attempting to calculate the constants by which the new R_2 must be multiplied for the generation of zeros in the a_{32} and a_{42} positions, the computer will essay division of $-\frac{8}{3}$ and $\frac{14}{3}$ by 0, and the program will fail. Fortunately, a simple refinement known as *partial pivoting* not only avoids this problem, but also helps to check any build-up of rounding errors in the course of the solution.

3.2.2 Gauss's Method with Partial Pivoting

Partial pivoting involves choosing for the pivot in each reduced matrix the element of largest magnitude in the relevant column, elements of rows which have previously been pivotal being excluded from consideration. In the following example, a broken line separates those rows which have already been pivotal from those which have not; the pivot (circled) is at each stage chosen from below this line. In order to obtain a final triangular form, the pivotal row must be interchanged with the row which would have been pivotal in the unmodified Gauss process.

$$\left(\begin{array}{cccc:c} \textcircled{3} & 1 & -2 & -1 & 3 \\ 2 & -2 & 2 & 3 & -8 \\ 1 & 5 & -4 & -1 & 3 \\ 3 & 1 & 2 & 3 & -1 \end{array}\right), \quad \begin{array}{l} R_1 \\ R_2 - \frac{2}{3}R_1 \\ R_3 - \frac{1}{3}R_2 \\ R_4 - R_1 \end{array} \left(\begin{array}{cccc:c} 3 & 1 & -2 & -1 & 3 \\ 0 & -\frac{8}{3} & \frac{10}{3} & \frac{11}{3} & -10 \\ 0 & \textcircled{\frac{14}{3}} & -\frac{10}{3} & -\frac{2}{3} & 2 \\ 0 & 0 & 4 & 4 & -4 \end{array}\right),$$

$$\begin{array}{l} R_1 \\ R_3 \\ R_2 + \frac{4}{7}R_3 \\ R_4 \end{array} \left(\begin{array}{cccc:c} 3 & 1 & -2 & -1 & 3 \\ 0 & \frac{14}{3} & -\frac{10}{3} & -\frac{2}{3} & 2 \\ 0 & 0 & \frac{10}{7} & \frac{23}{7} & -\frac{62}{7} \\ 0 & 0 & \textcircled{4} & 4 & -4 \end{array}\right), \quad \begin{array}{l} R_1 \\ R_2 \\ R_4 \\ R_3 - \frac{5}{14}R_4 \end{array} \left(\begin{array}{cccc:c} 3 & 1 & -2 & -1 & 3 \\ 0 & \frac{14}{3} & -\frac{10}{3} & -\frac{2}{3} & 2 \\ 0 & 0 & 4 & 4 & -4 \\ 0 & 0 & 0 & \textcircled{\frac{13}{7}} & -\frac{52}{7} \end{array}\right).$$

The solution, of course, is the same as before. This minor modification guarantees that Gauss's method will produce a solution, unless it is found at some stage that all those elements from amongst which the next pivot must be chosen are zero. In this case, the matrix of coefficients is singular (to the accuracy to which the machine is working, at least) and no unique solution exists.

A further advantage of partial pivoting is that the pivotal row is now always multiplied by a number of magnitude less than unity before its subtraction from other rows in the elimination process. Any rounding errors present in the pivotal row are thereby decreased in absolute magnitude, the propagated effect of these errors being decreased correspondingly.

Unlike the basic Gauss process, the modified version involves row interchanges, each of which reverses the sign of the determinant of the current reduced matrix. Since the pivots all appear on the diagonal of the final triangular matrix, we therefore have for the determinant of the original matrix

$$\Delta = \text{product of pivots} \times (-1)^m,$$

where m is the number of row interchanges. In the present example $m = 2$, and hence

$$\Delta = 3 \times \frac{14}{3} \times 4 \times \frac{13}{7} \times (-1)^2 = 104,$$

as we found previously.

In a practical computer program, the row interchanges referred to above may be accomplished by either of two means. The rows may actually be interchanged in the computer store, so that the final reduced matrix is triangular, as in the example given. Alternatively, the computer may be made to remember which row was used in the elimination of each variable, while the rows are kept in their original order. In this case, the final reduced matrix is not triangular, and the order in which the rows were used must be reversed in the back-substitution.

3.2.3 Gauss's Method with Full Pivoting

A logical extension of the idea of partial pivoting is known as *full* or *complete* pivoting. The columns are now not dealt with in their natural order, the pivot at each stage of the reduction being chosen as the element of largest magnitude in the submatrix of rows which have not hitherto been pivotal. The final reduced matrix is generally not triangular, even if rows are interchanged during the process, unless column interchanges are also made. In this case, the variables no longer occur in their original order in the equations which correspond to the matrix rows. In the following example, row interchanges alone are made.

$$\left(\begin{array}{cccc:c} 3 & 1 & -2 & -1 & 3 \\ 2 & -2 & 2 & 3 & -8 \\ 1 & \textcircled{5} & -4 & -1 & 3 \\ 3 & 1 & 2 & 3 & -1 \end{array}\right), \quad \begin{array}{l} R_3 \\ R_2 + \frac{2}{5}R_3 \\ R_1 - \frac{1}{5}R_3 \\ R_4 - \frac{1}{5}R_3 \end{array} \left(\begin{array}{cccc:c} 1 & 5 & -4 & -1 & 3 \\ \frac{12}{5} & 0 & \frac{2}{5} & \frac{13}{5} & -\frac{34}{5} \\ \frac{14}{5} & 0 & -\frac{6}{5} & -\frac{4}{5} & \frac{12}{5} \\ \frac{14}{5} & 0 & \frac{14}{5} & \textcircled{\frac{16}{5}} & -\frac{8}{5} \end{array}\right)$$

$$\begin{array}{l} R_1 \\ R_4 \\ R_3 + \frac{1}{4}R_4 \\ R_2 - \frac{13}{16}R_4 \end{array} \left(\begin{array}{cccc:c} 1 & 5 & -4 & -1 & 3 \\ \frac{14}{5} & 0 & \frac{14}{5} & \frac{16}{5} & -\frac{8}{5} \\ \textcircled{\frac{7}{2}} & 0 & -\frac{1}{2} & 0 & 2 \\ \frac{1}{8} & 0 & -\frac{15}{8} & 0 & -\frac{44}{8} \end{array}\right), \quad \begin{array}{l} R_1 \\ R_2 \\ R_3 \\ R_4 - \frac{1}{28}R_3 \end{array} \left(\begin{array}{cccc:c} 1 & 5 & -4 & -1 & 3 \\ \frac{14}{5} & 0 & \frac{14}{5} & \frac{16}{5} & -\frac{8}{5} \\ \frac{7}{2} & 0 & -\frac{1}{2} & 0 & 2 \\ 0 & 0 & \textcircled{-\frac{13}{7}} & 0 & -\frac{39}{7} \end{array}\right).$$

The back-substitution now gives the components of the solution in the order x_3, x_1, x_4, x_2. The final matrix may be made triangular by permuting the columns in the order 2413. This involves three interchanges (C_1 with C_2, C_2 with C_4 and C_3 with C_4); the elimination process also involved two row interchanges, and the determinant of the original matrix is therefore given by

$$\Delta = 5 \times \frac{16}{5} \times \frac{7}{2} \times -\frac{13}{7} \times (-1)^5 = 104,$$

since the pivots all lie on the diagonal of the triangular matrix.

Some disadvantages of full pivoting are the greater organisational problems, the longer search for the largest available pivot, and the fact that matrices of certain important particular forms, containing a high proportion of zero elements arranged in a special pattern, may have this pattern preserved by partial pivoting but destroyed by full pivoting (see Section 3.2.7). On the credit side, Wilkinson [5] has shown that the propagation of rounding errors is theoretically much better suppressed than by partial pivoting, though in practice no consistent significant increase in accuracy seems to result. It may be concluded that the simpler partial pivoting strategy is to be preferred for most purposes.

3.2.4 Jordan's Method

Jordan's method leads to a final reduced matrix whose left-hand part is diagonal rather than triangular. This form is achieved by using the pivotal row to generate zeros at every position in the relevant column except that occupied by the pivot itself. The following example illustrates the use of partial pivoting and row interchanges with Jordan's method:

$$\left(\begin{array}{cccc:c} \textcircled{3} & 1 & -2 & -1 & 3 \\ 2 & -2 & 2 & 3 & -8 \\ 1 & 5 & -4 & -1 & 3 \\ 3 & 1 & 2 & 3 & -1 \end{array}\right), \quad \begin{array}{l} R_1 \\ R_2 - \frac{2}{3}R_1 \\ R_3 - \frac{1}{3}R_1 \\ R_4 - R_1 \end{array} \left(\begin{array}{cccc:c} 3 & 1 & -2 & -1 & 3 \\ 0 & -\frac{8}{3} & \frac{10}{3} & \frac{11}{3} & -10 \\ 0 & \textcircled{\frac{14}{3}} & -\frac{10}{3} & -\frac{2}{3} & 2 \\ 0 & 0 & 4 & 4 & -4 \end{array}\right),$$

$$\begin{array}{l} R_1 - \frac{3}{14}R_3 \\ R_3 \\ R_2 + \frac{4}{7}R_3 \\ R_4 \end{array} \left(\begin{array}{cccc:c} 3 & 0 & -\frac{9}{7} & -\frac{6}{7} & \frac{18}{7} \\ 0 & \frac{14}{3} & -\frac{10}{3} & -\frac{2}{3} & 2 \\ 0 & 0 & \frac{10}{7} & \frac{23}{7} & -\frac{62}{7} \\ 0 & 0 & \textcircled{4} & 4 & -4 \end{array}\right), \quad \begin{array}{l} R_1 + \frac{9}{28}R_4 \\ R_2 + \frac{5}{6}R_4 \\ R_4 \\ R_3 - \frac{5}{14}R_4 \end{array} \left(\begin{array}{cccc:c} 3 & 0 & 0 & \frac{3}{7} & \frac{9}{7} \\ 0 & \frac{14}{3} & 0 & \frac{8}{3} & -\frac{4}{3} \\ 0 & 0 & 4 & 4 & -4 \\ 0 & 0 & 0 & \textcircled{\frac{13}{7}} & -\frac{52}{7} \end{array}\right)$$

$$\begin{array}{l} R_1 - \frac{3}{13}R_4 \\ R_2 - \frac{56}{39}R_4 \\ R_3 - \frac{28}{13}R_4 \\ R_4 \end{array} \left(\begin{array}{cccc:c} 3 & 0 & 0 & 0 & 3 \\ 0 & \frac{14}{3} & 0 & 0 & \frac{28}{3} \\ 0 & 0 & 4 & 0 & 12 \\ 0 & 0 & 0 & \frac{13}{7} & -\frac{52}{7} \end{array}\right), \quad \begin{array}{l} \frac{1}{3}R_2 \\ \frac{3}{14}R_2 \\ \frac{1}{4}R_3 \\ \frac{7}{13}R_4 \end{array} \left(\begin{array}{cccc:c} 1 & 0 & 0 & 0 & 1 \\ 0 & 1 & 0 & 0 & 2 \\ 0 & 0 & 1 & 0 & 3 \\ 0 & 0 & 0 & 1 & -4 \end{array}\right).$$

Again, the pivots have been circled. The final reduction of the left-hand part to unit matrix form exhibits the solution components explicitly on the right; no back-substitution is necessary, therefore, though the elimination process clearly requires more arithmetic than that of the Gauss processes. In fact, detailed analysis shows that the total arithmetic necessary for the solution of a single system of equations is larger by a factor of about $\frac{3}{2}$ for Jordan's method than for Gauss's method. Further, there is no simple pivotal strategy for Jordan's method which ensures that all row multipliers are less than one (note the multiplier $\frac{28}{13}$ in the fourth stage of the reduction above), and error analysis is much more difficult than for Gauss's method. However, both methods prove to be equally computationally efficient for inverting a matrix, and Jordan's is a convenient pencil-and-paper method in this context, though it is not recommended for use with a computer.

3.2.5 Matrix Inversion

Solution of a set of linear equations amounts to finding, for a given square matrix A and column vector $\mathbf{b}$, a column vector $\mathbf{x}$ satisfying $A\mathbf{x} = \mathbf{b}$. The related inversion problem is that of finding A^{-1} such that $AA^{-1} = I$, where A^{-1} and I have the same dimensions as A, I being a unit matrix. To take a 3 x 3 example, we have

$$\begin{pmatrix} a_{11} & a_{12} & a_{13} \\ a_{21} & a_{22} & a_{23} \\ a_{31} & a_{32} & a_{33} \end{pmatrix} \begin{pmatrix} \alpha_{11} & \alpha_{12} & \alpha_{13} \\ \alpha_{21} & \alpha_{22} & \alpha_{23} \\ \alpha_{31} & \alpha_{32} & \alpha_{33} \end{pmatrix} = \begin{pmatrix} 1 & 0 & 0 \\ 0 & 1 & 0 \\ 0 & 0 & 1 \end{pmatrix},$$

in which the a_{ij} are all known and we wish to determine the α_{ij}. The rules of matrix multiplication imply that the elements of the first column of the right-hand matrix depend upon the elements of all three rows of A taken in turn, but upon the elements of only the first column of A^{-1}. We may therefore calculate $(\alpha_{11}, \alpha_{21}, \alpha_{31})'$ by taking $(1, 0, 0)'$ as the right-hand side of a set of three linear equations. Similarly, we may calculate the elements of the second and third columns of A^{-1} by taking respectively the second and third columns of the unit matrix as our right-hand vector. The inversion of an $n \times n$ matrix is therefore seen to be equivalent to the solution of a system of n linear equations for n different right-hand sides, the columns of the appropriate unit matrix. Since the matrix of coefficients is the same in each case, we may deal with all the different right-hand sides simultaneously, as in the example which follows, where Jordan's method is used with natural pivoting:

$$\left(\begin{array}{cccc:cccc} \textcircled{3} & 1 & -2 & -1 & 1 & 0 & 0 & 0 \\ 2 & -2 & 2 & 3 & 0 & 1 & 0 & 0 \\ 1 & 5 & -4 & -1 & 0 & 0 & 1 & 0 \\ 3 & 1 & 2 & 3 & 0 & 0 & 0 & 1 \end{array}\right),$$

$$
\begin{array}{l}
R_1 \\[6pt] R_2 - \frac{2}{3}R_1 \\[6pt] R_3 - \frac{1}{3}R_1 \\[6pt] R_4 - R_1
\end{array}
\left(\begin{array}{cccc|cccc}
3 & 1 & -2 & -1 & 1 & 0 & 0 & 0 \\[6pt]
0 & -\frac{8}{3} & \frac{10}{3} & \frac{11}{3} & -\frac{2}{3} & 1 & 0 & 0 \\[6pt]
0 & \frac{14}{3} & -\frac{10}{3} & -\frac{2}{3} & -\frac{1}{3} & 0 & 1 & 0 \\[6pt]
0 & 0 & 4 & 4 & -1 & 0 & 0 & 1
\end{array}\right),
$$

$$
\begin{array}{l}
R_1 + \frac{3}{8}R_2 \\[6pt] R_2 \\[6pt] R_3 + \frac{7}{4}R_2 \\[6pt] R_4
\end{array}
\left(\begin{array}{cccc|cccc}
3 & 0 & -\frac{3}{4} & \frac{3}{8} & \frac{3}{4} & \frac{3}{8} & 0 & 0 \\[6pt]
0 & -\frac{8}{3} & \frac{10}{3} & \frac{11}{3} & -\frac{2}{3} & 1 & 0 & 0 \\[6pt]
0 & 0 & \frac{5}{2} & \frac{23}{4} & -\frac{3}{2} & \frac{7}{4} & 1 & 0 \\[6pt]
0 & 0 & 4 & 4 & -1 & 0 & 0 & 1
\end{array}\right),
$$

$$
\begin{array}{l}
R_1 + \frac{3}{10}R_3 \\[6pt] R_2 - \frac{4}{3}R_3 \\[6pt] R_3 \\[6pt] R_4 - \frac{8}{5}R_3
\end{array}
\left(\begin{array}{cccc|cccc}
3 & 0 & 0 & \frac{21}{10} & \frac{3}{10} & \frac{9}{10} & \frac{3}{10} & 0 \\[6pt]
0 & -\frac{8}{3} & 0 & -4 & \frac{4}{3} & -\frac{4}{3} & -\frac{4}{3} & 0 \\[6pt]
0 & 0 & \frac{5}{2} & \frac{23}{4} & -\frac{3}{2} & \frac{7}{4} & 1 & 0 \\[6pt]
0 & 0 & 0 & -\frac{26}{5} & \frac{7}{5} & -\frac{14}{5} & -\frac{8}{5} & 1
\end{array}\right),
$$

$$
\begin{array}{l}
R_1 + \frac{21}{52}R_4 \\[6pt] R_2 - \frac{10}{13}R_4 \\[6pt] R_3 + \frac{115}{104}R_4 \\[6pt] R_4
\end{array}
\left(\begin{array}{cccc|cccc}
3 & 0 & 0 & 0 & \frac{45}{52} & -\frac{3}{13} & -\frac{9}{26} & \frac{21}{52} \\[6pt]
0 & -\frac{8}{3} & 0 & 0 & \frac{10}{39} & \frac{32}{39} & -\frac{4}{39} & -\frac{10}{13} \\[6pt]
0 & 0 & \frac{5}{2} & 0 & \frac{5}{104} & -\frac{35}{26} & -\frac{10}{13} & \frac{115}{104} \\[6pt]
0 & 0 & 0 & -\frac{26}{5} & \frac{7}{5} & -\frac{14}{5} & -\frac{8}{5} & 1
\end{array}\right),
$$

$$\begin{matrix} \frac{1}{3}R_1 \\[2ex] \frac{3}{14}R_2 \\[2ex] \frac{1}{4}R_3 \\[2ex] \frac{7}{13}R_4 \end{matrix} \left(\begin{array}{cccc:cccc} 1 & 0 & 0 & 0 & \frac{15}{52} & -\frac{1}{13} & -\frac{3}{26} & \frac{7}{52} \\[1ex] 0 & 1 & 0 & 0 & -\frac{5}{52} & -\frac{4}{13} & \frac{1}{26} & \frac{15}{52} \\[1ex] 0 & 0 & 1 & 0 & \frac{1}{52} & -\frac{7}{13} & -\frac{4}{13} & \frac{23}{52} \\[1ex] 0 & 0 & 0 & 1 & -\frac{7}{26} & \frac{7}{13} & \frac{4}{13} & -\frac{5}{26} \end{array}\right).$$

The four columns on the right of this augmented matrix constitute the inverse of the original matrix. The Gauss process might equally well be used, but the elimination process does not then give the inverse explicitly, a separate back-substitution being required for each column of the inverse. For this reason, Gauss's method loses ground by comparison with Jordan's as far as the number of arithmetical operations is concerned, and the total arithmetic necessary to invert an $n \times n$ matrix proves to be identical for either method. Gauss's method is the more difficult to program, though it permits easier analysis of rounding errors.

It is simple to verify that the solution of the system of equations of the earlier example is

$$A^{-1}\begin{pmatrix} 3 \\ -8 \\ 3 \\ -1 \end{pmatrix} = \begin{pmatrix} 1 \\ 2 \\ 3 \\ -4 \end{pmatrix},$$

where A^{-1} is the inverse calculated above, since $A\mathbf{x} = \mathbf{b}$ may also be expressed as $\mathbf{x} = A^{-1}\mathbf{b}$. However, since inversion of an $n \times n$ matrix is equivalent to solving n systems of equations, it is obviously inefficient to solve a single system of equations by first evaluating the inverse of its coefficient matrix.

3.2.6 Application to Systems with Symmetric Matrices

The use of the basic Gauss process with a symmetric matrix permits a useful saving in computational effort, since the square submatrix of non-zero elements lying below those rows which have been pivotal is itself symmetric at each stage of the elimination. It is therefore only necessary to calculate the elements of this sub-matrix lying on or below the diagonal, the symmetry property being used to find the values of the remaining elements. Consider, in illustration, the following 4×4 system:

$$
\begin{pmatrix} 1 & 2 & 3 & 4 & \vdots & 5 \\ 2 & 3 & 4 & 1 & \vdots & 6 \\ 3 & 4 & 1 & 2 & \vdots & 7 \\ 4 & 1 & 2 & 3 & \vdots & 8 \end{pmatrix},
\quad
\begin{array}{l} R_1 \\ R_2 - 2R_1 \\ R_3 - 3R_1 \\ R_4 - 4R_1 \end{array}
\begin{pmatrix} 1 & 2 & 3 & 4 & \vdots & 5 \\ 0 & -1 & -2 & -7 & \vdots & -4 \\ 0 & -2 & -8 & -10 & \vdots & -8 \\ 0 & -7 & -10 & -13 & \vdots & -12 \end{pmatrix},
$$

$$
\begin{array}{l} R_1 \\ R_2 \\ R_3 - 2R_2 \\ R_4 - 7R_2 \end{array}
\begin{pmatrix} 1 & 2 & 3 & 4 & \vdots & 5 \\ 0 & -1 & -2 & -7 & \vdots & -4 \\ 0 & 0 & -4 & 4 & \vdots & 0 \\ 0 & 0 & 4 & 36 & \vdots & 16 \end{pmatrix},
\quad
\begin{array}{l} R_1 \\ R_2 \\ R_3 \\ R_4 + R_3 \end{array}
\begin{pmatrix} 1 & 2 & 3 & 4 & \vdots & 5 \\ 0 & -1 & -2 & -7 & \vdots & -4 \\ 0 & 0 & -4 & 4 & \vdots & 0 \\ 0 & 0 & 0 & 40 & \vdots & 16 \end{pmatrix}.
$$

The solution is $x_4 = x_3 = x_2 = \frac{2}{5}$, $x_1 = \frac{7}{5}$. The symmetric submatrices arising in this example have been enclosed by broken lines.

By taking advantage of the symmetry of the submatrices, we may reduce the total arithmetic involved in the solution of a large symmetric system by nearly one-half. Unfortunately, however, the basic Gauss process fails if a zero pivot occurs, and the use of any pivotal strategy to avoid this possibility in general destroys the symmetry of the submatrices. The important exceptional case is that of the positive definite symmetric matrix, for which it is possible to show that all the pivots in the basic Gauss process must be greater than zero. Furthermore, if full pivoting is used with a positive definite matrix, the pivotal element in any reduced matrix will be found to lie on the diagonal. The appropriate row interchange, *followed by the corresponding column interchange*, will then always result in a symmetric submatrix in the next reduced matrix. The following simple example illustrates the procedure:

$$
\begin{pmatrix} 2 & 1 & 0 & \vdots & 4 \\ 1 & 4 & 1 & \vdots & 12 \\ 0 & 1 & 3 & \vdots & 11 \end{pmatrix},
\quad
\begin{array}{l} R_2 \\ R_1 \\ R_3 \end{array}
\begin{pmatrix} 1 & 4 & 1 & \vdots & 12 \\ 2 & 1 & 0 & \vdots & 4 \\ 0 & 1 & 3 & \vdots & 11 \end{pmatrix},
\quad
\begin{array}{c} C_2 \quad C_1 \quad C_3 \\ \begin{pmatrix} 4 & 1 & 1 & \vdots & 12 \\ 1 & 2 & 0 & \vdots & 4 \\ 1 & 0 & 3 & \vdots & 11 \end{pmatrix} \end{array}.
$$

Here the first pivot has been chosen as the largest element in the array of coefficients; interchange of R_1 and R_2, followed by the corresponding interchange of C_1 and C_2, brings this element to the a_{11} position of a rearranged symmetric matrix. The first stage of the elimination now gives

$$
\begin{array}{l} R_1 \\ R_2 - \frac{1}{4}R_1 \\ R_3 - \frac{1}{4}R_1 \end{array}
\begin{pmatrix} 4 & 1 & 1 & \vdots & 12 \\ 0 & \frac{7}{4} & -\frac{1}{4} & \vdots & 1 \\ 0 & -\frac{1}{4} & \frac{11}{4} & \vdots & 8 \end{pmatrix},
\quad
\begin{array}{l} R_1 \\ R_3 \\ R_2 \end{array}
\begin{pmatrix} 4 & 1 & 1 & \vdots & 12 \\ 0 & -\frac{1}{4} & \frac{11}{4} & \vdots & 8 \\ 0 & \frac{7}{4} & -\frac{1}{4} & \vdots & 1 \end{pmatrix},
\quad
\begin{array}{c} C_1 \quad C_3 \quad C_2 \\ \begin{pmatrix} 4 & 1 & 1 & \vdots & 12 \\ 0 & \frac{11}{4} & -\frac{1}{4} & \vdots & 8 \\ 0 & -\frac{1}{4} & \frac{7}{4} & \vdots & 1 \end{pmatrix} \end{array},
$$

where the next pivot has now been chosen and manoeuvred into the top left-hand position of the 2 x 2 symmetric submatrix. And so the process continues. The next stage of the elimination gives an upper triangular matrix on the left, and the subsequent back-substitution gives the solution as $x_1 = 1, x_3 = 3, x_2 = 2$, the components being found in this order because of the column interchanges which took place.

3.2.7 Application to Systems with Tridiagonal Matrices

A matrix is said to be *banded* if all the elements in each row are zero except for the diagonal element and a limited number on either side of it. An important special case is the *tridiagonal* matrix, in which only the diagonal element and the elements immediately to its left and right in each row are non-zero. Such matrices commonly arise during the numerical solution of ordinary and partial differential equations, and in this context they are often also positive definite, in which case the basic Gauss process cannot fail due to occurrence of zero pivots. Consider the situation at the start of the third stage of elimination in an $n \times n$ case:

$$\left(\begin{array}{cccccccc:c}
* & * & & & & & & & * \\
0 & * & * & & & & & & * \\
 & 0 & * & * & & & & & * \\
 & & * & * & * & & & & * \\
 & & & * & * & * & & & * \\
 & & & & \cdot & \cdot & \cdot & \cdot & \cdot \\
 & & & & & * & * & * & * \\
 & & & & & & * & * & *
\end{array}\right).$$

Here non-zero elements are indicated by asterisks, and the zeros produced by previous steps are indicated. Since each modified row in the matrix contains only two non-zero elements, the $(k-1)^{\text{th}}$ such row is equivalent to a relation of the type

$$x_{k-1} + p_{k-1}x_k = q_{k-1}.$$

The kth row, which we will suppose to be as yet unmodified, represents the kth original equation:

$$a_k x_{k-1} + b_k x_k + c_k x_{k+1} = d_k.$$

Elimination of x_{k-1} between these equations gives

$$x_k + p_k x_{k+1} = q_k, \tag{2}$$

where
$$p_k = \frac{c_k}{b_k - a_k p_{k-1}} \tag{3}$$

and
$$q_k = \frac{d_k - a_k q_{k-1}}{b_k - a_k p_{k-1}}. \tag{4}$$

Thus the assumption of a linear relationship between x_{k-1} and x_k implies that x_k and x_{k+1} are similarly related. If we define $p_1 = c_1/b_1$ and $q_1 = d_1/b_1$ (so that equation (2) with $k = 1$ is equivalent to the first equation in the system), we may use equations (3) and (4) to generate $p_2, q_2, p_3, q_3, \ldots, p_n, q_n$. Since all the p's and q's are now known, we may set $x_{n+1} = 0$ (this is a "fictitious" variable) and use equation (2) to calculate, successively, $x_n, x_{n-1}, \ldots, x_1$. Clearly, the calculation of the p's and q's is equivalent to the elimination stage of Gauss's process, and the subsequent calculation of the x's to the back-substitution. For a tridiagonal matrix, then, the Gauss process can be concisely expressed in terms of recurrence relations, which is a very convenient feature from the computational point of view.

3.2.8 Application to Complex Systems of Equations

Complex systems may be solved by any of the standard methods if complex arithmetic is used. Alternatively, real arithmetic may be used throughout if advantage is taken of the fact that

$$I = \begin{pmatrix} 1 & 0 \\ 0 & 1 \end{pmatrix} \quad \text{and} \quad J = \begin{pmatrix} 0 & 1 \\ -1 & 0 \end{pmatrix}$$

have similar algebraic properties to 1 and i, the real and imaginary units. A complex number $a + ib$ may therefore be represented as

$$\begin{pmatrix} a & b \\ -b & a \end{pmatrix}.$$

If every element of an $n \times n$ complex matrix is replaced by such a 2×2 matrix, the result is a real $2n \times 2n$ matrix, which may be dealt with by any appropriate method of this chapter. In the following example, Jordan elimination has been used, with natural pivoting; the final result is therefore obtained without need for back-substitution:

$$(1 + i)z_1 - 2iz_2 = -5 + 3i$$
$$(3 - 2i)z_1 + (2 + i)z_2 = 18 - 5i.$$

We quote only the initial and final augmented matrices, omitting the intermediate details:

$$\left(\begin{array}{cccc:cc} 1 & 1 & 0 & -2 & -5 & 3 \\ -1 & 1 & 2 & 0 & -3 & -5 \\ 3 & -2 & 2 & 1 & 18 & -5 \\ 2 & 3 & -1 & 2 & 5 & 18 \end{array}\right), \ldots, \left(\begin{array}{cccc:cc} 1 & 0 & 0 & 0 & 3 & 2 \\ 0 & 1 & 0 & 0 & -2 & 3 \\ 0 & 0 & 1 & 0 & 1 & -3 \\ 0 & 0 & 0 & 1 & 3 & 1 \end{array}\right).$$

The solution is clearly $z_1 = 3 + 2i$, $z_2 = 1 - 3i$, and it may be noted that the second of the columns of constants is redundant, since this solution may be deduced from the first column alone on the right of the final matrix.

3.3 Matrix Analysis of Elimination Processes

We consider first the application of the basic Gauss method to the system of equations $A\mathbf{x} = \mathbf{b}$, or

$$\begin{pmatrix} a_{11} & a_{12} & \cdots & a_{1n} \\ a_{21} & a_{22} & \cdots & a_{2n} \\ \vdots & \vdots & & \vdots \\ a_{n1} & a_{n2} & \cdots & a_{nn} \end{pmatrix} \begin{pmatrix} x_1 \\ x_2 \\ \vdots \\ x_n \end{pmatrix} = \begin{pmatrix} b_1 \\ b_2 \\ \vdots \\ b_n \end{pmatrix}.$$

In the first stage of the elimination, the row operation $R_k - m_{k1}R_1$, where $m_{k1} = a_{k1}/a_{11}$ and $k = 2, 3, \ldots, n$, is used to generate the zeros required in the first column. It is easy to verify that the resulting system of reduced equations is $M_1A\mathbf{x} = M_1\mathbf{b}$, where

$$M_1 = \begin{pmatrix} 1 & & & & \\ -m_{21} & 1 & & & \\ -m_{31} & 0 & 1 & & \\ \vdots & \vdots & \vdots & \ddots & \\ -m_{n1} & 0 & 0 & \ldots\ 0 & 1 \end{pmatrix}.$$

Similarly, the second reduced system has the matrix representation $M_2M_1A\mathbf{x} = M_2M_1\mathbf{b}$, where

$$M_2 = \begin{pmatrix} 1 & & & & \\ 0 & 1 & & & \\ 0 & -m_{32} & 1 & & \\ \vdots & \vdots & \vdots & \ddots & \\ 0 & -m_{n2} & 0 & \ldots\ 0 & 1 \end{pmatrix},$$

in which the m_{k2} are the row multipliers of the second elimination stage. A continuation of this argument shows that the final reduced system is

$$MA\mathbf{x} = M\mathbf{b},$$

where $M = M_{n-1} \ldots M_2M_1$ is a product of lower triangular matrices (having only zero elements above their diagonals), and is therefore itself lower triangular.

But the end product of the elimination is a set of equations whose coefficient matrix U is *upper* triangular (all elements *below* the diagonal being zero). Thus we may write $MA = U$, or, on defining $M^{-1} = L$,

$$A = LU.$$

Here L, being the inverse of a lower triangle, is itself lower triangular. We conclude that, barring mishaps, a matrix A can be factorised as the product of a lower and an upper triangular matrix. Such a factorisation is impossible only for the case when one of the multipliers m_{ij} is infinite, corresponding to the occurrence of a zero pivot in the reduction process. This LU factorisation is the basis of the *triangular decomposition* methods for solving linear systems.

It is a remarkable fact that the elements of the lower triangular factor L are simply the row multipliers m_{ij} of the elimination process. For, since $M = M_{n-1} \ldots M_2 M_1$, we have $L = M^{-1} = M_1^{-1} M_2^{-1} \ldots M_{n-1}^{-1}$; now M_k^{-1} proves to be simply M_k itself with the signs of its off-diagonal elements reversed, and the product of all the M_k^{-1} in the indicated order is found to give

$$L = \begin{pmatrix} 1 & & & & & \\ m_{21} & 1 & & & & \\ m_{31} & m_{32} & 1 & & & \\ \vdots & \vdots & \vdots & \ddots & & \\ \vdots & \vdots & \vdots & & \ddots & \\ m_{n1} & m_{n2} & m_{n3} & \ldots & m_{n,n-1} & 1 \end{pmatrix}.$$

The reader is invided to confirm the details of this analysis for, say, the 4 x 4 case.

To illustrate the LU factorisation, we may consider the system dealt with in Section 3.2.1, whose coefficient matrix can be factorised as

$$\begin{pmatrix} 3 & 1 & -2 & -1 \\ 2 & -2 & 2 & 3 \\ 1 & 5 & -4 & -1 \\ 3 & 1 & 2 & 3 \end{pmatrix} = \begin{pmatrix} 1 & 0 & 0 & 0 \\ \frac{2}{3} & 1 & 0 & 0 \\ \frac{1}{3} & -\frac{7}{4} & 1 & 0 \\ 1 & 0 & \frac{8}{5} & 1 \end{pmatrix} \begin{pmatrix} 3 & 1 & -2 & -1 \\ 0 & -\frac{8}{3} & \frac{10}{3} & \frac{11}{3} \\ 0 & 0 & \frac{5}{2} & \frac{23}{4} \\ 0 & 0 & 0 & -\frac{26}{5} \end{pmatrix}.$$

We observe that the row-multipliers of the basic Gauss process all appear under the diagonal of L, and that U is the final reduced matrix of the process (see p. 44).

If partial pivoting is employed, then the rows are generally not used in their natural order. If we had prior knowledge of their order of use (which, in fact, we have not), we could rearrange them into this order before commencing the solution; the pivot at each stage would then fall on the diagonal, as in the basic Gauss process, and no row interchanges would be necessary during the reduction to triangular form. Thus Gauss's method with partial pivoting, applied to a system with matrix A, is equivalent to the basic Gauss process applied to some row-permutation of A, which will not have the same LU factorisation as A itself. The situation for complete pivoting is similar but more complicated, since we are now in effect applying the basic Gauss process to a row *and column* permutation of the matrix A.

Jordan's method with natural choice of pivots, applied to $A\mathbf{x} = \mathbf{b}$, is equivalent to $JA\mathbf{x} = D\mathbf{x} = J\mathbf{b}$, where $JA = D$ is a diagonal matrix. Here J is the product of matrices J_k, typified by

$$J_3 = \begin{pmatrix} 1 & 0 & m_{13} & 0 & \dots & 0 & 0 \\ 0 & 1 & m_{23} & 0 & \dots & 0 & 0 \\ 0 & 0 & 1 & 0 & \dots & 0 & 0 \\ 0 & 0 & m_{43} & 1 & \dots & 0 & 0 \\ & & & & & \vdots & \vdots \\ \vdots & \vdots & \vdots & \vdots & & 1 & 0 \\ 0 & 0 & m_{n3} & 0 & \dots & 0 & 1 \end{pmatrix},$$

and the product of the J_k has no special form of which advantage may be taken. There is no decomposition process, therefore, associated with Jordan elimination.

3.4 Triangular Decomposition Methods

3.4.1 Doolittle's Method

The LU factorisation of a matrix A is a by-product of Gaussian elimination; the row-multipliers m_{ij} give the off-diagonal elements of L, whose diagonal elements are all unity, while the final reduced matrix is U. The Gauss process comprises a number of elimination stages, each involving the calculation and recording of a new reduced matrix. The factorisation may be obtained more directly by Doolittle's method, which we illustrate for a general 4 x 4 matrix. In the relation $A = LU$, or

$$\begin{pmatrix} a_{11} & a_{12} & a_{13} & a_{14} \\ a_{21} & a_{22} & a_{23} & a_{24} \\ a_{31} & a_{32} & a_{33} & a_{34} \\ a_{41} & a_{42} & a_{43} & a_{44} \end{pmatrix} = \begin{pmatrix} 1 & 0 & 0 & 0 \\ m_{21} & 1 & 0 & 0 \\ m_{31} & m_{32} & 1 & 0 \\ m_{41} & m_{42} & m_{43} & 1 \end{pmatrix} \begin{pmatrix} u_{11} & u_{12} & u_{13} & u_{14} \\ 0 & u_{22} & u_{23} & u_{24} \\ 0 & 0 & u_{33} & u_{34} \\ 0 & 0 & 0 & u_{44} \end{pmatrix},$$

all the m's and u's are initially unknown. The rule for matrix multiplication enables them to be found from the following equations:

(i) from the first row of A,

$$\underline{u_{11}} = a_{11}, \quad \underline{u_{12}} = a_{12}, \quad \underline{u_{13}} = a_{13}, \quad \underline{u_{14}} = a_{14},$$

(ii) from the first column of A,

$$\underline{m_{21}} u_{11} = a_{21}, \quad \underline{m_{31}} u_{11} = a_{31}, \quad \underline{m_{41}} u_{11} = a_{41},$$

(iii) from the second row of A,

$$m_{21} u_{12} + \underline{u_{22}} = a_{22}, \quad m_{21} u_{13} + \underline{u_{23}} = a_{23}, \quad m_{21} u_{14} + \underline{u_{24}} = a_{24},$$

(iv) from the second column of A,

$$m_{31} u_{12} + \underline{m_{32}} u_{22} = a_{32}, \qquad m_{41} u_{12} + \underline{m_{42}} u_{22} = a_{42},$$

(v) from the third row of A,

$$m_{31} u_{13} + m_{32} u_{23} + \underline{u_{33}} = a_{33}, \qquad m_{31} u_{14} + m_{32} u_{24} + \underline{u_{34}} = a_{34},$$

(vi) from the third column of A,

$$m_{41} u_{13} + m_{42} u_{23} + \underline{m_{43}} u_{33} = a_{43},$$

(vii) finally, from the fourth row of A,

$$m_{41} u_{14} + m_{42} u_{24} + m_{43} u_{34} + \underline{u_{44}} = a_{44}.$$

If they are used in the order given, each of these equations contains only one unknown variable (underlined). The process can clearly be extended to the factorisation of matrices of any order. For the $n \times n$ case, it may be summarised as

$$\left.\begin{aligned} &\text{for } j = i, i+1, \ldots, n: \quad u_{ij} = a_{ij} - \sum_{k=1}^{i-1} m_{ik} u_{kj} \\ &\text{for } j = i+1, i+2, \ldots, n: \quad m_{ji} = \frac{1}{u_{ii}} \left[a_{ji} - \sum_{k=1}^{i-1} m_{jk} u_{ki} \right] \end{aligned}\right\} \quad i = 1, 2, \ldots, n. \tag{5}$$

This factorisation method fails only if one of the diagonal elements of U, which are used as divisors in the second of the relations above, proves to be zero. This corresponds to the occurrence of a zero pivot in Gauss's process (it may be remembered that all the pivots appear on the diagonal of the final reduced matrix in that process). We shall shortly outline a row-interchanging modification of Doolittle factorisation which overcomes this problem.

Once L and U have been determined, the system $A\mathbf{x} = \mathbf{b}$ can be re-expressed as the pair of equations

$$L\mathbf{y} = \mathbf{b}$$

$$\text{and} \quad U\mathbf{x} = \mathbf{y},$$

which are both very easy to solve since their coefficient matrices are triangular.

Investigation shows that the total burden of arithmetic is the same as for Gauss's method, which is hardly surprising, since the two methods are obviously closely related.

The disadvantage of Gauss's process in comparison with Doolittle's is that each time an element of a reduced matrix is computed and stored, a rounding error generally occurs. Doolittle's process permits these errors to be largely evaded by the use of double-precision arithmetic in the calculation of the elements of L and U from equations (5). The results may then be rounded to single precision and recorded on the completion of each calculation. The removal of the need for computing and recording several intermediate matrices has, therefore, localised what might otherwise be a significant source of error to a single step in the determination of each element of L and U. The use of double precision arithmetic in this step leads to a degree of accuracy comparable with that attained if the entire Gauss process were carried out with double precision. This latter is an unattractive proposition, in that it would require twice as much computer storage as the corresponding single-precision solution.

3.4.2 Doolittle's Method with Row Interchanges

We have seen that Doolittle factorisation of a matrix A may fail on the occurrence of a zero diagonal element of U, corresponding to the occurrence of a zero pivot in Gaussian elimination. We may avoid this possibility by introducing a modification analogous to the use of partial pivoting. Now, each time one of the diagonal elements of U is evaluated, the effect of row interchanges is examined, with the object of finding the arrangement which maximises the absolute value of the element concerned.

As shown in Section 3.3, the use of partial pivoting with Gauss's method in solving $A\mathbf{x} = \mathbf{b}$ is equivalent to the use of the basic method on $A^*\mathbf{x} = \mathbf{b}^*$, where A^* and $\mathbf{b}^*$ are corresponding row-permutations of A and $\mathbf{b}$. Then we wish our modified process to be equivalent to

$$MA^*\mathbf{x} = U\mathbf{x} = M\mathbf{b}^*.$$

On premultiplication by $M^{-1} = L$, this becomes

$$L\mathbf{y} = \mathbf{b}^*,$$

where

$$U\mathbf{x} = \mathbf{y},$$

and it is this pair of triangular systems to which we wish to reduce the original square system. Since $MA^* = U$ and $M\mathbf{b}^* = \mathbf{y}$, we may obtain U and $\mathbf{y}$ simultaneously by the same sequence of row operations. This we do in the following example, the appropriate row-permutations being determined as the solution progresses. The system of equations is that of Section 3.2.1:

$$\overset{A}{\left(\begin{array}{cccc} 3 & 1 & -2 & -1 \\ 2 & -2 & 2 & 3 \\ 1 & 5 & -4 & -1 \\ 3 & 1 & 2 & 3 \end{array}\right.}\ \vdots\ \overset{\mathbf{b}}{\left.\begin{array}{c} 3 \\ -8 \\ 3 \\ -1 \end{array}\right)} = \overset{[L]}{\begin{pmatrix} 1 & 0 & 0 & 0 \\ * & 1 & 0 & 0 \\ * & * & 1 & 0 \\ * & * & * & 1 \end{pmatrix}} \overset{[U]}{\left(\begin{array}{cccc} * & * & * & * \\ 0 & * & * & * \\ 0 & 0 & * & * \\ 0 & 0 & 0 & * \end{array}\right.}\ \vdots\ \overset{[\mathbf{y}]}{\left.\begin{array}{c} * \\ * \\ * \\ * \end{array}\right)}.$$

Here $[L]$, $[U]$ and $[\mathbf{y}]$ are bracketed to indicate the fact that our eventual L, U and $\mathbf{y}$ will correspond, not to A and $\mathbf{b}$, but to A^* and $\mathbf{b}^*$.

Initially, our only information is that the diagonal elements of L are all unity. If we leave $(A, \mathbf{b})$ unchanged, then $u_{11} = a_{11} = 3$. Interchange of R_2, R_3 or R_4 with R_1 gives u_{11} the values 2, 1, or 3 respectively. Since there is no advantage in interchanging at this stage, we leave R_1 of $(A, \mathbf{b})$ where it is; this enables us to calculate the elements of R_1 of $(U, \mathbf{y})$ and those of C_1 of L. We then have

$$\left(\begin{array}{cccc} 3 & 1 & -2 & -1 \\ 2 & -2 & 2 & 3 \\ 1 & 5 & -4 & -1 \\ 3 & 1 & 2 & 3 \end{array}\ \vdots\ \begin{array}{c} 3 \\ -8 \\ 3 \\ -1 \end{array}\right) = \begin{pmatrix} 1 & 0 & 0 & 0 \\ \frac{2}{3} & 1 & 0 & 0 \\ \frac{1}{3} & * & 1 & 0 \\ 1 & * & * & 1 \end{pmatrix} \left(\begin{array}{cccc} 3 & 1 & -2 & -1 \\ 0 & * & * & * \\ 0 & 0 & * & * \\ 0 & 0 & 0 & * \end{array}\ \vdots\ \begin{array}{c} 3 \\ * \\ * \\ * \end{array}\right).$$

Next, we consider the element u_{22}. If, on the left, R_2 remains in its present position, then $u_{22} = a_{22} - m_{21}u_{12} = -\frac{8}{3}$ (the m's, remember, are elements of L). Interchange of R_3 or R_4 with R_2 on the left would give $u_{22} = \frac{14}{3}$ or $u_{22} = 0$ respectively. We accordingly interchange R_2 and R_3 in order to obtain the maximum possible magnitude for u_{22}. The elements of C_1 of L must be similarly interchanged. The remaining elements in R_2 of $(U, \mathbf{y})$ and in C_2 of L may now be computed, the result being

$$\left(\begin{array}{cccc} 3 & 1 & -2 & -1 \\ 1 & 5 & -4 & -1 \\ 2 & -2 & 2 & 3 \\ 3 & 1 & 2 & 3 \end{array}\ \vdots\ \begin{array}{c} 3 \\ 3 \\ -8 \\ -1 \end{array}\right) = \begin{pmatrix} 1 & 0 & 0 & 0 \\ \frac{1}{3} & 1 & 0 & 0 \\ \frac{2}{3} & -\frac{4}{7} & 1 & 0 \\ 1 & 0 & * & 1 \end{pmatrix} \left(\begin{array}{cccc} 3 & 1 & -2 & -1 \\ 0 & \frac{14}{3} & -\frac{10}{3} & -\frac{2}{3} \\ 0 & 0 & * & * \\ 0 & 0 & 0 & * \end{array}\ \vdots\ \begin{array}{c} 3 \\ 2 \\ * \\ * \end{array}\right).$$

We may now either leave R_3 on the left where it is, or interchange it with R_4. The former alternative gives $u_{33} = a_{33} - m_{31}u_{13} - m_{32}u_{23} = \frac{10}{7}$, while the latter gives $u_{33} = 4$, which is the larger of the two available values. After the interchange, u_{34}, y_3 and m_{43} may be calculated. Lastly, since the row-permutation of A is now completely determined, u_{44} and y_4 may also be found. We then have the final result

$$\overset{A^*}{\begin{pmatrix} 3 & 1 & -2 & -1 & \vdots & 3 \\ 1 & 5 & -4 & -1 & \vdots & 3 \\ 3 & 1 & 2 & 3 & \vdots & -1 \\ 2 & -2 & 2 & 3 & \vdots & -8 \end{pmatrix}} \overset{\mathbf{b}^*}{} = \overset{L}{\begin{pmatrix} 1 & 0 & 0 & 0 \\ \frac{1}{3} & 1 & 0 & 0 \\ 1 & 0 & 1 & 0 \\ \frac{2}{3} & -\frac{4}{7} & \frac{5}{14} & 1 \end{pmatrix}} \overset{U \qquad\qquad \mathbf{y}}{\begin{pmatrix} 3 & 1 & -2 & -1 & \vdots & 3 \\ 0 & \frac{14}{3} & -\frac{10}{3} & -\frac{2}{3} & \vdots & 2 \\ 0 & 0 & 4 & 4 & \vdots & -4 \\ 0 & 0 & 0 & \frac{13}{7} & \vdots & -\frac{52}{7} \end{pmatrix}}.$$

The solution $\mathbf{x}$ is now found from the triangular system $U\mathbf{x} = \mathbf{y}$ by back-substitution in the usual manner. Note that the matrix U is the same as was obtained using Gaussian elimination with partial pivoting in Section 3.2.2. Its diagonal elements (the pivots in the latter process) are all $\geqslant 1$ in magnitude, and the growth of rounding errors is inhibited accordingly. Thus the use of row interchanges with the Doolittle method enables us to combine the virtues of partial pivoting with further enhancement of accuracy when double-precision arithmetic is used for calculating the elements of L and U.

At first sight, it appears that the introduction of row interchanges involves us in more arithmetic, since several possible values of each diagonal element of U must be computed and compared in magnitude. However, the calculation involved in obtaining the rejected values is not wasted, since if these values are divided by the chosen diagonal element the results are the elements of the next column of L to be determined. Apart from the search process, then, the total arithmetic is the same either with or without row interchanges (and, as pointed out earlier, the same as for Gauss's method).

From the comments in the two preceding paragraphs, it is apparent that Doolittle triangular decomposition with row interchanges is the best method for general-purpose use of those discussed here, provided a computer is available which permits the use of double-precision arithmetic. Otherwise, there is little to choose between the elimination and decomposition methods.

3.4.3 Matrix Inversion by Triangular Decomposition

As explained in Section 3.2.5, the inversion of an $n \times n$ matrix A involves the solution of $A\mathbf{x} = \mathbf{b}$ for n different right-hand sides, namely the columns of the appropriate unit matrix. Methods involving triangular decomposition are no less suitable than elimination methods in this context, and they are capable of giving improved accuracy, at little extra expense of computing time, if double-precision arithmetic is used as explained earlier.

The most straightforward approach is to obtain the LU decomposition of A as shown in the last section and then solve successively $LL^{-1} = I$ and $UA^{-1} = L^{-1}$. Some other inversion methods connected with triangular decomposition are mentioned by Fox [3], but they appear to have no particular computational advantage. For the particular case of a symmetric positive-definite matrix, the method of Cholsky (see Section 3.4.5.) has several advantages.

3.4.4 Alternative LU Factorisations

The factorisation of a matrix A as the product of lower and upper triangular matrices is by no means unique. In fact, the diagonal elements of one or other of the factors may be chosen arbitrarily; all the remaining elements of the upper and lower triangles may then be uniquely determined as in Doolittle's method, which is the case arising when all the diagonal elements of L are chosen to be unity. The name of Crout is often associated with triangular decomposition methods, and in Crout's variant the diagonal elements of U are all taken as unity. Apart from this, there is little distinction, as regards either procedure or accuracy, between the two methods. The reader may like to undertake for himself the Crout factorisation of the matrix of Section 3.2.1.

It is suggested by Wilkinson [4] (p. 117) that a choice of diagonal elements such that $m_{rr} \simeq |u_{rr}|$ may give better accuracy than either the Doolittle or Crout decomposition. This proposal is particularly relevant in connection with symmetric matrices, as we see in the next section.

3.4.5 Cholesky's Method for Symmetric Matrices

It is often possible to find an LU decomposition of a symmetric matrix A such that $U = L'$. We shall denote the particular matrix L having this property by Λ, so that $A = \Lambda\Lambda'$. As an illustration, consider the decomposition of a general 3×3 symmetric matrix in the form

$$\underset{A}{\begin{pmatrix} a_{11} & a_{12} & a_{13} \\ a_{21} & a_{22} & a_{23} \\ a_{31} & a_{32} & a_{33} \end{pmatrix}} = \underset{\Lambda}{\begin{pmatrix} \lambda_{11} & 0 & 0 \\ \lambda_{21} & \lambda_{22} & 0 \\ \lambda_{31} & \lambda_{32} & \lambda_{33} \end{pmatrix}} \underset{\Lambda'}{\begin{pmatrix} \lambda_{11} & \lambda_{21} & \lambda_{31} \\ 0 & \lambda_{22} & \lambda_{32} \\ 0 & 0 & \lambda_{33} \end{pmatrix}}.$$

The λ's may be determined from the following equations:

(i) from the first row of A,

$$\underline{\lambda}_{11}^2 = a_{11}, \quad \lambda_{11}\underline{\lambda}_{21} = a_{12}, \quad \lambda_{11}\underline{\lambda}_{31} = a_{13},$$

(ii) from the second row of A,

$$\lambda_{21}^2 + \underline{\lambda}_{22}^2 = a_{22}, \quad \lambda_{21}\lambda_{31} + \lambda_{22}\underline{\lambda}_{32} = a_{23},$$

(iii) from the third row of A,

$$\lambda_{31}^2 + \lambda_{32}^2 + \underline{\lambda}_{33}^2 = a_{33}.$$

The unknown in each equation has been underlined. Note that equations need only be written for the elements of A lying on and above the diagonal; because of the symmetry of A we have $a_{ji} = a_{ij}$, and the corresponding equations for the elements of A lying below the diagonal merely duplicate those written above. For an $n \times n$ matrix, the equations for Cholesky decomposition may be written as

$$\left.\begin{aligned} &\text{for } i = j: && \lambda_{ii} = \left[a_{jj} - \sum_{k=1}^{j-1} \lambda^2{}_{jk}\right]^{\frac{1}{2}} \\ &\text{for } i = j+1, j+2, \ldots, n: && \lambda_{ij} = \frac{1}{\lambda_{jj}}\left[a_{ji} - \sum_{k=1}^{j-1} \lambda_{ik}\,\lambda_{jk}\right] \end{aligned}\right\} j = 1, 2, \ldots, n.$$

There is clearly an advantage here over Crout or Doolittle decomposition, in that once $L = \Lambda$ is found, $U = \Lambda'$ follows immediately. This is the reason why a separate equation corresponding to each element of A is unnecessary. For a large matrix, the computational effort required for a Cholesky decomposition is little more than half that needed otherwise.

The method is not without its snags, however. Firstly it may be found that one or more of the diagonal elements of Λ is the square root of a negative number. This does not invalidate the factorisation, but it does mean that other pure imaginary elements will arise during the computation. No complex numbers will occur, however, and a little ingenuity can avoid the need for use of complex arithmetic. Secondly, the process fails if any diagonal element of Λ is zero, and in general the use of row interchanges to avoid this destroys the symmetry of the system. For this reason, Cholesky decomposition is perhaps best reserved for use with positive-definite matrices, for which case it may be shown that the diagonal elements of Λ are all real and greater than zero. Further, for the positive definite case the equivalent of full pivoting may be used, since the appropriate row and column interchanges now retain the symmetry of the system. A useful gain in accuracy is thereby achieved. The resulting method may be expected to be slightly slower than the elimination method of the second example in Section 3.2.6, owing to the necessity for extracting square roots. However, detailed analysis shows the Cholesky method to have a slight advantage in accuracy (see Section 3.4.4), and double-precision arithmetic may be used in computing the elements of Λ to gain a further improvement. Thus, Cholesky's method is an attractive proposition for problems involving symmetric positive-definite matrices.

As an example, we will take the positive-definite problem of Section 3.2.6. The solution given there is equivalent to a Doolittle decomposition with row and column interchanges; the corresponding Cholesky solution commences as follows:

$$\begin{array}{ccc} A \quad\quad\quad b & [\Lambda] & [\Lambda'] \quad\quad\quad \mathbf{y} \\ \begin{pmatrix} 2 & 1 & 0 & \vdots & 4 \\ 1 & 4 & 1 & \vdots & 12 \\ 0 & 1 & 3 & \vdots & 11 \end{pmatrix} = & \begin{pmatrix} * & 0 & 0 \\ * & * & 0 \\ * & * & * \end{pmatrix} & \begin{pmatrix} * & * & * & \vdots & * \\ 0 & * & * & \vdots & * \\ 0 & 0 & * & \vdots & * \end{pmatrix}. \\ x_1 \; x_2 \; x_3 & & x_1 \; x_2 \; x_3 \end{array}$$

Now $\lambda_1^2 = a_{11}$, and in order to give $|\lambda_1|$ the maximum possible value we must manoeuvre the largest element of A to the a_{11} position. This involves the interchanges of R_1 and R_2 and of C_1 and C_2. From the first row which results on the left, we may calculate $\lambda_{11}, \lambda_{21}, \lambda_{31}$ and y_1:

$$\begin{pmatrix} 4 & 1 & 1 & \vdots & 12 \\ 1 & 2 & 0 & \vdots & 4 \\ 1 & 0 & 3 & \vdots & 11 \end{pmatrix} = \begin{pmatrix} 2 & 0 & 0 \\ \frac{1}{2} & * & 0 \\ \frac{1}{2} & * & * \end{pmatrix} \begin{pmatrix} 2 & \frac{1}{2} & \frac{1}{2} & \vdots & 6 \\ 0 & * & * & \vdots & * \\ 0 & 0 & * & \vdots & * \end{pmatrix}.$$
$$\quad x_2 \;\; x_1 \;\; x_3 \qquad\qquad\qquad\qquad\qquad x_2 \;\; x_1 \;\; x_3$$

Note that a column interchange on the left alters the significance of the columns of Λ' correspondingly. Now a_{22} is given by $\lambda_{21}^2 + \lambda_{22}^2$, and we must choose a_{22} from between the two remaining diagonal elements on the left to give λ_{22} its maximum value. With no reordering, we obtain $\lambda_{22} = \frac{1}{2}\sqrt{7}$. Interchange of R_2 and R_3, C_2 and C_3 (which entails also the corresponding row and column interchanges, respectively, of the established elements of Λ and Λ'), gives $\lambda_{22} = \frac{1}{2}\sqrt{11}$, and since this is the larger of the two values we make the interchange. At this stage, the row and column permutation of A is fixed, and we may evaluate all the remaining elements of Λ, Λ', $\mathbf{y}$:

$$\overset{A^*\qquad\quad \mathbf{b}^*}{\begin{pmatrix} 4 & 1 & 1 & \vdots & 12 \\ 1 & 3 & 0 & \vdots & 11 \\ 1 & 0 & 2 & \vdots & 4 \end{pmatrix}} = \overset{\Lambda}{\begin{pmatrix} 2 & 0 & 0 \\ \frac{1}{2} & \frac{\sqrt{11}}{2} & 0 \\ \frac{1}{2} & -\frac{1}{2\sqrt{11}} & \sqrt{\frac{19}{11}} \end{pmatrix}} \overset{\Lambda' \qquad\qquad\qquad \mathbf{y}}{\begin{pmatrix} 2 & \frac{1}{2} & \frac{1}{2} & \vdots & 6 \\ 0 & \frac{\sqrt{11}}{2} & -\frac{1}{2\sqrt{11}} & \vdots & \frac{16}{\sqrt{11}} \\ 0 & 0 & \sqrt{\frac{19}{11}} & \vdots & \sqrt{\frac{19}{11}} \end{pmatrix}}.$$
$$x_2 \;\; x_3 \;\; x_1 \qquad\qquad\qquad\qquad\qquad\qquad x_2 \;\; x_3 \qquad\quad x_1$$

The solution is now found from the augmented matrix $(\Lambda', \mathbf{y})$ to be $x_1 = 1$, $x_3 = 3, x_2 = 2$.

3.5 Ill-conditioned Systems

3.5.1 The Problem of Ill-conditioning

A matrix is singular if one or more of its rows is linearly dependent on the others. If A is singular, then $A\mathbf{x} = \mathbf{b}$ has no solution in general, though for certain vectors $\mathbf{b}$ solutions exist which are not unique.

If A has only *approximate* linear dependence amongst its rows, $A\mathbf{x} = \mathbf{b}$ will have a unique solution, though certain difficulties may arise in finding it. A matrix suffering from this state of near-singularity is said to be *ill-conditioned.* This term was defined more precisely in Chapter 2.

If we attempt to solve a singular system $A\mathbf{x} = \mathbf{b}$ by Gaussian elimination using exact arithmetic, the process will fail because eventually the group of elements from amongst which the next pivot must be chosen will all be zero. This will happen whatever pivotal strategy is used. Practical computer arithmetic, however, introduces rounding errors which are likely to make these elements non-zero, though of small magnitude. Thus the computer may convert an original singular system into a non-singular system, and may then go on to produce a 'solution'. When A is not singular but is ill-conditioned, a similar situation may arise, in that eventually a pivot must be chosen from amongst a group of elements of small magnitude. A frequent symptom of ill-conditioning, then, is a large ratio of the magnitudes of the largest and smallest pivots used in the solution.*

Another feature of an ill-conditioned system $A\mathbf{x} = \mathbf{b}$ is that a small change in any element of A may occasion a very large change in the solution. To take a very simple example, consider

$$x_1 + 2x_2 = 2$$
$$(1 - \epsilon)x_1 + 2x_2 = 3.$$

When $\epsilon = 0$, the system is obviously singular, and there is no solution. For $\epsilon \neq 0$ we have $x_1 = -\epsilon^{-1}, x_2 = \frac{1}{2}(1 + 2\epsilon)\epsilon^{-1}$, and hence $\epsilon = 0.001$ gives $x_1 = -1000$, $x_2 = 501$, while $\epsilon = -0.001$ gives $x_1 = 1000, x_2 = -499$. Thus a change of -0.002 in one coefficient changes the solution component x_1 by 2000.

In view of such behaviour, we see that the rounding errors introduced during the computer solution of an ill-conditioned system, though small in themselves, may have a very significant effect on the accuracy of the computed solution. The importance of using pivotal strategy to prevent their undue growth must therefore be re-emphasised. If, in spite of all precautions, the true and computed solutions still differ appreciably, the technique next discussed may be used to improve matters.

3.5.2 Improvement of Solutions by Residual Correction

Suppose that $\mathbf{x}^{(0)}$ is a computed solution of $A\mathbf{x} = \mathbf{b}$. Then we can calculate the *residual vector* $\mathbf{b} - A\mathbf{x}^{(0)} = \boldsymbol{\eta}^{(0)}$, which gives some indication as to how closely $\mathbf{x}^{(0)}$ satisfies the equation. On adding these two relations and defining $\mathbf{x} - \mathbf{x}^{(0)} = \mathbf{d}$, we obtain

$$A\mathbf{d} = \boldsymbol{\eta}^{(0)}.$$

Thus the correction vector $\mathbf{d}$ satisfies another equation with matrix A. If the LU decomposition of A is available from the earlier solution, this equation may quickly be solved by successive solution of the two triangular systems $L\mathbf{y} = \boldsymbol{\eta}^{(0)}$ and $U\mathbf{d} = \mathbf{y}$. This requires only a fraction of the work of the original solution. If L and U were exact and no further rounding errors arose, $\mathbf{x}^{(0)} + \mathbf{d}$ would then

* Though Wilkinson [5] (p. 254) shows that this particular sympton does not invariably occur.

be the exact solution. In practice, however, rounding errors only permit the calculation of $\mathbf{d}^{(0)}$, an approximation to $\mathbf{d}$. The best we can hope for, then, is that $\mathbf{x}^{(1)} = \mathbf{x}^{(0)} + \mathbf{d}^{(0)}$ is an improved approximation to $\mathbf{x}$. New residual and correction vectors $\boldsymbol{\eta}^{(1)}$ and $\mathbf{d}^{(1)}$ may now be calculated, and the process repeated in an iterative manner until no further improvement results. Normally, only a small number of improvement cycles are necessary.

An essential feature of this process is that the residual $\boldsymbol{\eta}^{(r)} = \mathbf{b} - A\mathbf{x}^{(r)}$ must always be calculated with double precision. Otherwise, if $\mathbf{x}^{(r)}$ is close to the true solution, $\mathbf{b}$ and $A\mathbf{x}^{(r)}$ may be so nearly equal that a drastic loss of significant figures occurs in forming their difference. Single precision suffices for the rest of the computation.

It is pointed out by Fox [3] (p. 144) that this improvement scheme may be summed up in the iterative relation $\mathbf{x}^{(r+1)} = \mathbf{x}^{(r)} + B^{-1}(\mathbf{b} - A\mathbf{x}^{(r)})$, where B^{-1} is an approximate inverse of A satisfying $\mathbf{x}^{(0)} = B^{-1}\mathbf{b}$. A corresponding procedure for improving a calculated inverse $C^{(0)}$ of A is $C^{(r+1)} = C^{(r)}(2I - AC^{(r)})$, which he shows to have very rapid convergence. Wilkinson [5] (p. 255 ff.) gives a theoretical discussion of the convergence criterion for the former process, though for present purposes we may assume that convergence will result except in pathologically ill-conditioned cases.

We conclude with an example. The system $A\mathbf{x} = \mathbf{b}$, where $\mathbf{b} = (3.7, 4.7, 6.5)'$ and A and its Doolittle LU factorisation are

$$\begin{pmatrix} 1.0 & 1.2 & 1.5 \\ 1.2 & 1.5 & 2.0 \\ 1.5 & 2.0 & 3.0 \end{pmatrix} = \begin{pmatrix} 1 & 0 & 0 \\ 1.2 & 1 & 0 \\ 1.5 & 3.33 & 1 \end{pmatrix} \begin{pmatrix} 1.0 & 1.2 & 1.5 \\ 0 & 0.06 & 0.2 \\ 0 & 0 & 0.084 \end{pmatrix},$$

has the exact solution $\mathbf{x} = (1, 1, 1)'$. Here the elements of L and U are quoted to 3 significant figures, having been calculated in double precision to 6 s.f. and subsequently rounded, as suggested in Section 3.4.1. On solving successively $L\mathbf{y} = \mathbf{b}$ and $U\mathbf{x} = \mathbf{y}$, using single-precision arithmetic, we obtain the rather inaccurate first solution $\mathbf{x}^{(0)} = (0.87, 1.17, 0.952)'$. A double-precision calculation of the residual vector gives $\boldsymbol{\eta}^{(0)} = (-0.002, -0.003, 0.001)'$, correct to 6 s.f.; the single-precision solution of $A\mathbf{d}^{(0)} = \boldsymbol{\eta}^{(0)}$ is then $\mathbf{d}^{(0)} = (0.131, -0.168, 0.0476'$, which gives $\mathbf{x}^{(1)} = \mathbf{x}^{(0)} + \mathbf{d}^{(0)} = (1.00, 1.00, 1.00)'$, to 3 s.f. Thus a single application of the correction process has given full single-precision accuracy.

3.6. Practical Considerations

3.6.1 Computer Storage Requirements

It is often expedient, in using either an elimination or a decomposition method, to store the elements of both L and U (as shown in Section 3.3, the elements of L are the row multipliers m_{ij} which arise in the elimination process). This is because many computations require the solution of a system of equations for a number of right-hand sides, not all of which are known initially. The residual correction process of Section 3.5.2 offers an example of this situation. With L and U available, a solution for any new right-hand side $\mathbf{b}$ is obtainable from the triangular systems $L\mathbf{y} = \mathbf{b}$ and $U\mathbf{x} = \mathbf{y}$, by forward and backward substitution respectively, in only a small proportion of the time required for a single solution from scratch. The intermediate solution $\mathbf{y}$ corresponds to the transformed column vector resulting when an elimination process is applied to the augmented matrix $(A, \mathbf{b})$ as in Sections 3.2.1–3.2.4.

Unless a copy of the original matrix A is required, it may be overwritten as the computation proceeds, since once an element of L or U has been calculated the corresponding element of A (or any of its reduced versions) is no longer required. Thus, to take a 4 x 4 case, the space originally occupied by A finally contains the array

$$\begin{pmatrix} u_{11} & u_{12} & u_{13} & u_{14} \\ m_{21} & u_{22} & u_{23} & u_{24} \\ m_{31} & m_{32} & u_{33} & u_{34} \\ m_{41} & m_{42} & m_{43} & u_{44} \end{pmatrix}.$$

The missing diagonal elements of L are all unity for the Gauss or Doolittle processes. As a refinement, the diagonal elements may be replaced by their reciprocals, since these can then be used multiplicatively in determining the m_{ij} and in the subsequent back-substitution, multiplication being a more efficient process for a computer than division. If A is overwritten, then, the storage needed for the solution of the $n \times n$ system $A\mathbf{x} = \mathbf{b}$ is that for $n(n+1)$ variables plus a little working space; the constant vector can be overwritten by the final solution. Note, however, that the residual correction procedure of Section 3.5.2 requires the retention of A for the calculation of residual vectors, which roughly doubles the storage requirements.

The inversion problem amounts to the solution of n linear equations for n right-hand sides. Storage is therefore needed for $2n^2$ variables if A is overwritten, and $3n^2$ if it is not, plus working space.

3.6.2 Arithmetical Labour of the Computation

It is shown by Fox [3] (Chapter 7) that the number of operations required to solve $A\mathbf{x} = \mathbf{b}$ and to invert A (a general $n \times n$ matrix) are as follows:

METHOD	SOLUTION			INVERSION		
	÷	×	+	÷	×	+
Gauss	n	$\frac{1}{3}n^3 + n^2 - \frac{1}{3}n$	$\frac{1}{3}n^3 + \frac{1}{2}n^2 - \frac{5}{6}n$	n	$n^3 - 1$	$n^3 - 2n^2 + n$
Decomposition	n	$\frac{1}{3}n^3 + n^2 - \frac{1}{3}n$	$\frac{1}{3}n^3 + \frac{1}{2}n^2 - \frac{5}{6}n$	n	$n^3 - 1$	$n^3 - 2n^2 + n$
Jordan	n	$\frac{1}{2}n^3 + n^2 - \frac{1}{2}n$	$\frac{1}{2}n^3 - \frac{1}{2}n$	n	$n^3 - 1$	$n^3 - 2n^2 + n$

From this point of view, all three methods are equally efficient for the inversion, though Jordan's is less so than the others for solving a single system of equations. If A is symmetric and positive definite, use of the symmetric version of Gauss's method (Section 3.2.6) or Cholesky's method (Section 3.4.5) gives a significant saving. Both methods require $0(\frac{1}{6}n^3)$ multiplications and $0(\frac{1}{6}n^3)$ additions; the first requires n reciprocals and the second n square roots. Thus Cholesky's method is likely to be slightly the less efficient arithmetically, though this is compensated for by its superior accuracy. The use of row, or row and column, interchanges with any of the methods involves no additional arithmetic, but extra time is needed for searching for pivots and organising the interchanges.

In any of these processes, the elimination or decomposition accounts for the majority of the work. If both L and U are stored, as suggested in Section 3.6.1, the solution for a new right-hand side requires only $0(n^2)$ further multiplications and additions.

3.6.3 Scaling and Equilibration

Consider a system $A\mathbf{x} = \mathbf{b}$, whose augmented matrix is $(A, \mathbf{b})$. Multiplication of any row of $(A, \mathbf{b})$ by a constant clearly does not affect the solution; multiplication of any column of A by a constant, however, has the effect of dividing the corresponding solution component by that constant. By means of such row and column scalings, we can obtain a new system $B\mathbf{y} = \mathbf{c}$, say, whose solution $\mathbf{y}$ is simply related to $\mathbf{x}$; the matrix B is said to be a *scaled equivalent* of A.

Any non-singular matrix A possesses a *condition number* (see Chapter 2), which is large when $A\mathbf{x} = \mathbf{b}$ is ill-conditioned and of order unity for a very well-conditioned system. Now if $A\mathbf{x} = \mathbf{b}$ is ill-conditioned, its scaled version $B\mathbf{y} = \mathbf{c}$ may be less so, in the sense that its condition number is smaller. Then the scaled system is more stable than the original as regards the growth of rounding errors, and we will probably obtain a more accurate version of $\mathbf{x}$ by solving $B\mathbf{y} = \mathbf{c}$ for $\mathbf{y}$ and reversing the effect of the column scaling, rather than by solving $A\mathbf{x} = \mathbf{b}$ directly. The improvement results from the different sequence of pivots arising in the solution of the scaled system.

Unfortunately, the subject of scaling is still not perfectly understood. In particular, the problem of finding, for a given matrix A, the scaled equivalent B with minimum possible condition number is not completely solved (but see Chapter 11 of Forsythe and Moler [1]). Wilkinson [5] (p. 129 ff.) claims that the scaling of each row and column so that its maximum element has magnitude between $\frac{1}{2}$ and 1 ($\frac{1}{4}$ and 1 in symmetric cases, if symmetry is to be preserved in the scaling) generally reduces the condition number of a matrix. He calls this process *equilibration*, and it is not difficult to perform by trial and error for small matrices, though no satisfactory general method for equilibrating large matrices has yet been devised. Indeed, equilibration as described above is not always possible, and when it is possible the set of scaling factors is not unique. Row-equilibration alone is less satisfactory but generally worthwhile, and Wilkinson [6] has shown that its equivalent may be incorporated in an equation-solving program so that no preliminary scaling need be made. Curtiss and Reid [2] propose a method for approximate row and column equilibration. Whatever method is used, the rows and columns should be scaled by powers of 2 so that (in the binary arithmetic of the computer) no rounding errors are introduced.

References

1. CURTISS, A. R. & REID, J. K., *J. Inst. Maths. Applics. 10*, 118 (1972).
2. FORSYTHE, G. & MOLER, C. B., *Computer Solution of Linear Algebraic Systems*, Prentice-Hall (1967).
3. FOX, L., *An Introduction to Numerical Linear Algebra*, O.U.P. (1964).
4. NATIONAL PHYSICAL LABORATORY, *Rounding Errors in Algebraic Processes*, No. 32 of *Notes on Applied Science*, H.M.S.O. (1963).
5. WILKINSON, J. H., *The Algebraic Eigenvalue Problem*, O.U.P. (1965).
6. WILKINSON, J. H., The Solution of Ill-conditioned Linear Equations, in *Mathematical Methods for Digital Computers, Vol. 2* (Ed. RALSTON, A. & WILF, H. S.), Wiley (1967).

Exercises

A desk or pocket calculator will be needed for several of these Exercises, particularly those marked with an asterisk.

1.* Solve the following system of equations by Gaussian elimination with (i) natural pivoting, (ii) partial pivoting and (iii) complete pivoting. Use 4-decimal floating-point arithmetic (i.e. round to four significant figures after every arithmetical operation). Compare the computed solutions with the exact solution $(\frac{1}{2}, \frac{1}{3}, \frac{1}{4})'$.

$$\begin{aligned} 0.3114x_1 + 0.9438x_2 + 0.4720x_3 &= 0.5883 \\ 0.3808x_1 + 0.7614x_2 + 0.5028x_3 &= 0.5699 \\ 0.5486x_1 + 0.9051x_2 + 0.6856x_3 &= 0.7474. \end{aligned}$$

2.* Obtain the Doolittle decomposition of the matrix of Exercise 1, again using 4-decimal floating-point arithmetic, and using row interchanges as detailed in Section 3.4.2. Hence calculate the solution, and show that agreement is obtained with (ii) above.

3.* Starting with the calculated solution of Exercise 2, use the method of residual correction to obtain the correct solution to four significant figures. Note that the residuals must be computed to eight significant figures.

4. Solve the system of equations

$$\begin{aligned}2x_1 - x_2 \qquad\qquad\qquad &= 1\\ -x_1 + 2x_2 - x_3 \qquad\quad &= 2\\ - x_2 + 2x_3 - x_4 &= 3\\ - x_3 + 2x_4 &= 4\end{aligned}$$

by the method of Section 3.2.7.

5. Obtain the Doolittle and Cholesky factorisations of the positive-definite matrix of Exercise 4.

6. Use Jordan's method, with natural pivoting and using exact arithmetic, to find the inverse of

$$A = \begin{pmatrix} 1 & 2 & 0 & 1 \\ -1 & -1 & 1 & 0 \\ 2 & 3 & 0 & 0 \\ 1 & 4 & -1 & 5 \end{pmatrix}.$$

7. Invert the matrix A of Exercise 6 using Jordan's method with partial pivoting, and working with 3-decimal floating-point arithmetic. Note the emergence of large row multipliers.

8.* Use Doolittle decomposition, with row interchanges, to invert the matrix A of Exercise 6, working with 3-decimal floating-point arithmetic.

9.* Use Cholesky's method, with row and column interchanges, to solve

$$\begin{aligned}0.5624x_1 - 0.3831x_2 + 0.1072x_2 &= 0.1803\\ 0.3176x_1 + 0.6834x_2 - 0.5016x_3 &= 0.2612\\ 0.1122x_1 - 0.4302x_2 + 0.9632x_3 &= 0.1535.\end{aligned}$$

Use 4-decimal floating-point arithmetic, and compare your results with the exact solution $(\frac{1}{2}, \frac{1}{3}, \frac{1}{4})'$.

10. (i) Show that, if A is a non-symmetric square matrix, then $A'A$ is symmetric. Show also that, if B is symmetric and non-singular, B^2 is positive-definite (hint: show that $\mathbf{x}'B^2\mathbf{x} > 0$, where $\mathbf{x}$ is an arbitrary vector).

(ii) Determine the number of arithmetical operations used in multiplying together two $n \times n$ matrices. Hence show that no computational advantage may be gained by transforming a non-symmetric system into a symmetric, or a symmetric system into a positive-definite, using the results of (i).

11. (i) Determine the number of multiplications and additions needed for the evaluation of an $n \times n$ determinant, both directly and by the method of Sections 3.2.1–3.2.3.

(ii) Cramer's Rule gives the solution components of $A\mathbf{x} = \mathbf{b}$ explicitly as $x_i = |A^{(i)}|/|A|$, where $A^{(i)}$ is the matrix A with its ith column replaced by the vector **b**. Show that a numerical solution using this result is less efficient arithmetically than Gaussian elimination.

12. Consider the system

$$\begin{aligned} x_1 + 2x_2 + 256x_3 &= 5 \\ 2x_1 + x_2 + 256x_3 &= 6 \\ x_1 + x_2 \qquad\quad &= 3. \end{aligned}$$

Show that row-scaling using the factors $2^{-8}, 2^{-8}, 1$ results in an equilibrated system, as does column-scaling using the factors $\frac{1}{2}, \frac{1}{2}, 2^{-8}$.

Solve the two scaled systems using 4-decimal floating-point arithmetic and partial pivoting. Note that the scaling affects the sequence of rows from which the pivots are chosen, and that one scaled system exhibits signs of ill-conditioning while the other does not.

CHAPTER 4

THE SOLUTION OF LINEAR SYSTEMS OF EQUATIONS BY ITERATIVE METHODS

4.1 Introduction

Three closely related methods are studied in this chapter, all *iterative* or *indirect* in nature. Unlike the direct methods of Chapter 3, which attempt to calculate an exact solution in a finite number of operations, these methods start with an initial approximation and generate successively improved approximations in an infinite sequence whose *limit* is the exact solution. In practical terms, this is no great disadvantage. The direct solution will be subject to rounding errors, while the indirect method can in principle approach the exact solution arbitrarily closely, given sufficient iterations of its refinement process. In some cases, the indirect method requires much less computation than the direct method to achieve a comparable accuracy in the solution.

4.2 Iterative Methods in Action

4.2.1 Jacobi's Method

The simplest of the methods we shall deal with is that attributed to *Jacobi*, sometimes known also as the *method of simultaneous displacements.* It is rarely used in practice, but is of some theoretical interest and provides a convenient starting-point for our study. The following 4 x 4 system of equations will serve to illustrate the method:

$$\begin{aligned} 2x_1 - x_2 \qquad\qquad &= 1 \\ -x_1 + 2x_2 - x_3 \qquad &= 2 \\ - x_2 + 2x_3 - x_4 &= 3 \\ - x_3 + 2x_4 &= 4. \end{aligned} \tag{1}$$

This system of equations, whose exact solution is

$$x_1 = 4, \quad x_2 = 7, \quad x_3 = 8, \quad x_4 = 6,$$

is *tridiagonal*, having non-zero elements only on the diagonal of its matrix of coefficients and immediately to either side of it in each row. Such systems commonly arise in the numerical treatment of differential equations.

The first step in the solution is to use the first equation to express x_1 in terms of the other variables, the second equation to express x_2, and so on. The resulting relations are

$$\begin{aligned} x_1 &= \tfrac{1}{2}(1 + x_2) \\ x_2 &= \tfrac{1}{2}(2 + x_1 + x_3) \\ x_3 &= \tfrac{1}{2}(3 + x_2 + x_4) \\ x_4 &= \tfrac{1}{2}(4 + x_3). \end{aligned} \tag{2}$$

The addition of some superscripts turns these into a set of recurrence relations:

$$\begin{aligned} x_1^{(r+1)} &= \tfrac{1}{2}(1 + x_2^{(r)}) \\ x_2^{(r+1)} &= \tfrac{1}{2}(2 + x_1^{(r)} + x_3^{(r)}) \\ x_3^{(r+1)} &= \tfrac{1}{2}(3 + x_2^{(r)} + x_4^{(r)}) \\ x_4^{(r+1)} &= \tfrac{1}{2}(4 + x_3^{(r)}). \end{aligned} \tag{3}$$

Here the superscript (r) labels a component (presumed known) of the approximate solution after r iterations. Having no prior idea as to the approximate values of the solution components, we will arbitrarily take as initial values

$$x_1^{(0)} = x_2^{(0)} = x_3^{(0)} = x_4^{(0)} = 0.$$

With these values put into their right-hand sides, equations (3) give

$$x_1^{(1)} = \tfrac{1}{2}, \quad x_2^{(1)} = 1, \quad x_3^{(1)} = \tfrac{3}{2}, \quad x_4^{(1)} = 2.$$

The next iteration starts with the new set of values, and so the process continues. The results obtained, rounded to two decimal places, are as follows:

r	$x_1^{(r)}$	$x_2^{(r)}$	$x_3^{(r)}$	$x_4^{(r)}$
0	0.00	0.00	0.00	0.00
1	0.50	1.00	1.50	2.00
2	1.00	2.00	3.00	2.75
3	1.50	3.00	3.88	3.50
4	2.00	3.69	4.75	3.94
5	2.34	4.38	5.31	4.38
6	2.69	4.83	5.88	4.66
7	2.91	5.28	6.24	4.94
8	3.14	5.58	6.61	5.12
9	3.29	5.88	6.85	5.30
10	3.44	6.07	7.09	5.42
11	3.53	6.26	7.25	5.54
12	3.63	6.39	7.40	5.62
13	3.70	6.52	7.51	5.70
14	3.76	6.60	7.61	5.75
15	3.80	6.68	7.68	5.81
16	3.84	6.74	7.74	5.84
17	3.87	6.79	7.79	5.87
18	3.90	6.83	7.83	5.89
19	3.91	6.86	7.86	5.92
20	3.93	6.89	7.89	5.93
21	3.94	6.91	7.91	5.95
22	3.96	6.93	7.93	5.95
23	3.96	6.94	7.94	5.96
24	3.97	6.95	7.95	5.97
25	3.98	6.96	7.96	5.98
26	3.98	6.97	7.97	5.98
27	3.98	6.98	7.97	5.98
28	3.99	6.98	7.98	5.99
29	3.99	6.98	7.98	5.99
30	3.99	6.99	7.99	5.99
31	3.99	6.99	7.99	5.99
32	3.99	6.99	7.99	5.99
33	4.00	6.99	7.99	6.00
34	4.00	6.99	7.99	6.00
35	4.00	7.00	8.00	6.00
36	4.00	7.00	8.00	6.00

We see that 35 iterations of Jacobi's method give the solution to two decimal place accuracy.

4.2.2 The Gauss-Seidel Method

A simple modification of Jacobi's process leads to the *Gauss-Seidel* method, which is much more of a practical proposition. It is often referred to as the *method of successive displacements* or (particularly in connection with the numerical solution of elliptic partial differential equations) *Liebmann's method.* Instead of waiting until each cycle of the iteration is complete before substituting the improved values into the recurrence relations, we now use the updated values as soon as they become available. If we consider once again the solution of equations (1), their initial rearrangement in the form of equations (2) is the same, but equations (3) are replaced by

$$\begin{aligned}
x_1^{(r+1)} &= \tfrac{1}{2}(1 + x_2^{(r)}) \\
x_2^{(r+1)} &= \tfrac{1}{2}(2 + x_1^{(r+1)} + x_3^{(r)}) \\
x_3^{(r+1)} &= \tfrac{1}{3}(3 + x_2^{(r+1)} + x_4^{(r)}) \\
x_4^{(r+1)} &= \tfrac{1}{2}(4 + x_3^{(r+1)}).
\end{aligned} \tag{4}$$

We see that once $x_1^{(r+1)}$ has been calculated from the first equation, it is *immediately* used in the second equation, and so on. If the same starting values are taken as previously, namely

$$x_1^{(0)} = x_2^{(0)} = x_3^{(0)} = x_4^{(0)} = 0,$$

the results are as follows, to two decimal places:

r	$x_1^{(r)}$	$x_2^{(r)}$	$x_3^{(r)}$	$x_4^{(r)}$	r	$x_1^{(r)}$	$x_2^{(r)}$	$x_3^{(r)}$	$x_4^{(r)}$
0	0.00	0.00	0.00	0.00	11	3.92	6.90	7.92	5.96
1	0.50	1.25	2.13	3.06	12	3.95	6.94	7.95	5.97
2	1.13	2.63	4.34	4.17	13	3.97	6.96	7.97	5.98
3	1.81	4.08	5.63	4.81	14	3.98	6.97	7.98	5.99
4	2.54	5.08	6.45	5.22	15	3.99	6.98	7.99	5.99
5	3.04	5.74	6.98	5.49	16	3.99	6.99	7.99	6.00
6	3.37	6.18	7.33	5.67	17	3.99	6.99	7.99	6.00
7	3.59	6.46	7.56	5.78	18	4.00	6.99	8.00	6.00
8	3.73	6.65	7.72	5.86	19	4.00	7.00	8.00	6.00
9	3.82	6.77	7.81	5.91	20	4.00	7.00	8.00	6.00
10	3.88	6.85	7.88	5.94					

In this example, the Gauss-Seidel method has given the solution correct to two decimal place accuracy after only 19 iterations, compared with the 35 needed by Jacobi's method. Both methods require exactly the same number of arithmetical operations per iterative cycle, and the Gauss-Seidel modification of Jacobi's process has therefore reduced the total computational labour by a factor of around 2. This figure is fairly typical. The Gauss-Seidel method is also superior to Jacobi's process in requiring less computer storage. Whereas the latter stores all the elements of $\mathbf{x}^{(r)}$ until all the elements of $\mathbf{x}^{(r+1)}$ have been computed, in the former the elements of $\mathbf{x}^{(r)}$ may be overwritten by the elements of $\mathbf{x}^{(r+1)}$ as they are calculated. Storage is therefore needed only for one vector rather than two. The next method to be illustrated has this same advantage.

4.2.3 Successive Over-relaxation

A further simple modification of our improved process leads to a method which is still more powerful, though it is not always easy to make full use of its potentialities. This new development is called *successive over-relaxation* (usually abbreviated as S.O.R.) or the *extrapolated Gauss-Seidel* or *extrapolated Liebmann* method.

The recurrence relations of the Gauss-Seidel process, equations (4), may be rewritten as

$$x_1^{(r+1)} = x_1^{(r)} + \tfrac{1}{2}(1 - 2x_1^{(r)} + x_2^{(r)})$$
$$x_2^{(r+1)} = x_2^{(r)} + \tfrac{1}{2}(2 + x_1^{(r+1)} - 2x_2^{(r)} + x_3^{(r)})$$
$$x_3^{(r+1)} = x_3^{(r)} + \tfrac{1}{2}(3 + x_2^{(r+1)} - 2x_3^{(r)} + x_4^{(r)})$$
$$x_4^{(r+1)} = x_4^{(r)} + \tfrac{1}{2}(4 + x_3^{(r+1)} - 2x_4^{(r)}).$$

The reader will observe that the terms in $x_i^{(r)}$ on the right-hand side of the ith equation may be cancelled, the results then being identical with equations (3). In this new form, the relations express $x_i^{(r+1)}$ as $x_i^{(r)}$ plus a correction term (*displacement term*). The S.O.R. method improves the convergence of the Gauss-Seidel method by increasing the displacement term by a factor ω, the *relaxation factor*, which usually lies between 1 and 2.* The iterative relations for S.O.R. are thus

$$x_1^{(r+1)} = x_1^{(r)} + \tfrac{1}{2}\omega(1 - 2x_1^{(r)} + x_3^{(r)})$$
$$x_2^{(r+1)} = x_2^{(r)} + \tfrac{1}{2}\omega(2 + x_1^{(r+1)} - 2x_2^{(r)} + x_3^{(r)})$$
$$x_3^{(r+1)} = x_3^{(r)} + \tfrac{1}{2}\omega(3 + x_2^{(r+1)} - 2x_3^{(r)} + x_4^{(r)})$$
$$x_4^{(r+1)} = x_4^{(r)} + \tfrac{1}{2}\omega(4 + x_3^{(r+1)} - 2x_4^{(r)}).$$

We will defer for the moment any discussion of the best value to take for ω. To illustrate the method we will take $\omega = 1.2$ and set

$$x_1^{(0)} = x_2^{(0)} = x_3^{(0)} = x_4^{(0)} = 0,$$

as previously. The results obtained are these:

r	$x_1^{(r)}$	$x_2^{(r)}$	$x_3^{(r)}$	$x_4^{(r)}$
0	0.00	0.00	0.00	0.00
1	0.60	1.56	2.74	4.04
2	1.42	3.38	5.71	5.01
3	2.34	5.35	6.88	5.53
4	3.34	6.26	7.50	5.79
5	3.69	6.66	7.77	5.90

r	$x_1^{(r)}$	$x_2^{(r)}$	$x_3^{(r)}$	$x_4^{(r)}$
6	3.86	6.85	7.90	5.96
7	3.94	6.93	7.95	5.98
8	3.97	6.97	7.98	5.99
9	3.99	6.99	7.99	6.00
10	3.99	6.99	8.00	6.00
11	4.00	7.00	8.00	6.00
12	4.00	7.00	8.00	6.00

Here we have attained 2 decimal place accuracy after only 11 iterations, and it is clear that S.O.R. is considerably more efficient, at least for this particular example, than the Gauss-Seidel method.

* Sometimes a value of ω in the range $0 < \omega < 1$ may be used, in which case the term *underrelaxation* is more appropriate.

4.2.4 Computational Features of Iterative Methods

Iterative methods may be used to greatest advantage when the matrix of coefficients of the equations to be solved is *sparse*, containing a high proportion of zero elements. In many such cases the use of an elimination method would have the effect of destroying the sparseness of the matrix by replacing zeros with non-zero elements, particularly above the leading diagonal. All these non-zero elements must, of course, be stored by the computer, and it is here that iterative methods gain, in that they continually operate with the original sparse matrix, of which only the small proportion of non-zero elements must be stored. Indeed, in some applications these elements can be *calculated* whenever they are needed, so that even less storage is required; we then refer to a *generated* (rather than a *stored*) matrix. The problem of Section 4.5 offers an example of this possibility. In either case, an iterative technique usually enables a given computer to solve larger sparse systems than it could cope with by an elimination method.

It is difficult to make a general statement concerning the computational labour necessary for solving a system of equations by an iterative method. When an elimination method is applied to an $n \times n$ system of equations, the number of multiplications and of additions required is roughly $\frac{1}{3}n^3$ for large n. It is easy to see that the number of both multiplications and additions needed by an iterative method is roughly n^2 per iteration for a general $n \times n$ system with no zero coefficients. Thus an iterative method may have the advantage from this point of view, provided the number of iterations needed to give the accuracy desired is less than about $\frac{1}{3}n$. Whether this is so depends upon the rate of convergence of the process (which is difficult to estimate *a priori* in many cases), and upon the availability of a good starting approximation. If the system of equations is sparse, however, the number of multiplications and additions performed per iteration may be very much smaller than n^2, in which case a much larger number of iterations can be tolerated.

Iterative methods, then, are not suitable for general-purpose use, but may be appropriate for solving $n \times n$ systems of equations having one of more of the following characteristics:

(i) n^2 is large compared with the computer core storage available,
(ii) the matrix of coefficients is sparse, the majority of its elements being zero.
(iii) a particularly good starting approximation is known.

In fact, these three criteria are often satisfied by the very large systems of linear equations which can arise in the numerical solution of ordinary or partial differential equations. Iterative methods are of paramount importance in this connection, since it is frequently impracticable to solve such systems by direct methods because of the prohibitive amount both of computer storage and of execution time involved.

A practical point which should be made is that, although for the sake of simplicity we have illustrated the application of iterative methods to a tridiagonal system of equations, systems of this particular form are normally more efficiently solved by the direct method of Section 3.2.7 of the last chapter. It is shown by Varga [4] (p. 195) that this method requires a total arithmetical effort roughly equivalent to only two iterations of S.O.R., which are hardly likely to result in comparable accuracy in the solution.

To conclude this appraisal, a cautionary note. The methods discussed here cannot, in general, be guaranteed to succeed. They suffer, in common with most iterative methods, from the possibility that convergence to the desired solution may not occur, and that the approximate solutions generated may progressively *diverge* from that solution. Fortunately, there are cases of frequent occurrence for which convergence is assured. In particular, at least one of the methods given here will converge when applied to a system of equations $A\mathbf{x} = \mathbf{b}$ such that either

(i) A is symmetric and positive-definite, or
(ii) A is *strictly diagonally dominant*, in the sense that the magnitude of each diagonal element of A is larger than the sum of the magnitudes of all the other elements in the same row.

Other less immediately useful sufficient conditions for convergence also exist, and we shall meet some of them later when we examine the question of convergence in some detail.

4.3 Matrix Analysis of Iterative Methods

4.3.1 Formulation in Matrix Terms

In this section, we take a rather general look at the iterative processes previously illustrated. In particular, we apply matrix methods to establish criteria for their convergence. We will suppose that we wish to solve the $n \times n$ system of equations

$$\begin{aligned} a_{11}x_1 + a_{12}x_2 + \ldots + a_{1n}x_n &= b_1 \\ a_{21}x_1 + a_{22}x_2 + \ldots + a_{2n}x_n &= b_2 \\ \cdots\cdots\cdots\cdots\cdots\cdots\cdots\cdots \\ a_{n1}x_1 + a_{n2}x_2 + \ldots + a_{nn}x_n &= b_n, \end{aligned}$$

which may be written in matrix form as

$$A\mathbf{x} = \mathbf{b}.$$

It is convenient to express A as the sum of the three matrices

$$L = \begin{pmatrix} 0 & 0 & \dots & 0 & 0 \\ a_{21} & 0 & \dots & 0 & 0 \\ a_{31} & a_{32} & \dots & 0 & 0 \\ \dots & \dots & \dots & \dots & \dots \\ a_{n1} & a_{n2} & \dots & a_{n,n-1} & 0 \end{pmatrix} \qquad U = \begin{pmatrix} 0 & a_{12} & \dots a_{1,n-1} & & a_{1n} \\ \dots & \dots & \dots & \dots & \dots \\ 0 & 0 & \dots & a_{n-2,n-1} & a_{n-2,n} \\ 0 & 0 & \dots & 0 & a_{n-1,n} \\ 0 & 0 & \dots & 0 & 0 \end{pmatrix}$$

and $D = \text{diag}(a_{11}, a_{22}, \dots, a_{nn})$, thus:

$$A = L + D + U.$$

Our system of equations may then be written as

$$(L + D + U)\mathbf{x} = \mathbf{b},$$

whence $$D\mathbf{x} = -L\mathbf{x} - U\mathbf{x} + \mathbf{b}$$

or $$\mathbf{x} = -D^{-1} L\mathbf{x} - D^{-1}U\mathbf{x} + D^{-1}\mathbf{b}. \tag{5}$$

Here it is assumed that D^{-1} exists, which, since $D^{-1} = \text{diag}\,(a_{11}^{-1}, a_{22}^{-1}, \dots, a_{nn}^{-1})$, means that none of the diagonal elements of the original matrix A can be zero. If zero diagonal elements do occur, the iterative processes will not work, though a simple re-ordering of the equations may suffice to remedy this.

Our three methods may all be derived from equation (5). The addition of superscripts, $(r + 1)$ on the left and (r) on the right, leads to the recurrence relation

$$\mathbf{x}^{(r+1)} = -D^{-1}(L + U)\mathbf{x}^{(r)} + D^{-1}\mathbf{b}. \tag{6}$$

This equation, written out in full, proves to be the rearrangement of the original system of equations appropriate for Jacobi iteration, in which all terms except those on the diagonal are taken over to the right-hand sides, and each resulting equation is then divided by the coefficient of the diagonal term which remains on the left. Alternatively, we may allocate different superscripts to the two terms in $\mathbf{x}$ on the right-hand side of equation (5), to obtain

$$\mathbf{x}^{(r+1)} = -D^{-1}L\mathbf{x}^{(r+1)} - D^{-1}U\mathbf{x}^{(r)} + D^{-1}\mathbf{b}, \tag{7}$$

which is the Gauss-Seidel rearrangement of the original system. To see this, we recall that in Gauss-Seidel iteration, when the value of $x_i^{(r+1)}$ is being calculated using the ith equation, the values of $x_1^{(r+1)}, x_2^{(r+1)}, \dots, x_{i-1}^{(r+1)}$ are already available and in use. These components have as their coefficients the elements of A lying beneath the diagonal, namely the non-zero elements of L. For components whose coefficients are the non-zero elements of U we must be content with using values from the previous iteration, $x_{i+1}^{(r)}, x_{i+2}^{(r)}, \dots, x_n^{(r)}$.

Our third method, S.O.R., modifies the displacement vector $\mathbf{d}^{(r)} = \mathbf{x}^{(r+1)} - \mathbf{x}^{(r)}$ of the Gauss-Seidel process by a factor ω. Equation (7) may be re-expressed as

$$\mathbf{x}^{(r+1)} = \mathbf{x}^{(r)} + \{D^{-1}\mathbf{b} - D^{-1}L\mathbf{x}^{(r+1)} - (I + D^{-1}U)\mathbf{x}^{(r)}\},$$

where the term in curly brackets is the displacement vector. Then the S.O.R. relation is

$$\mathbf{x}^{(r+1)} = \mathbf{x}^{(r)} + \omega\{D^{-1}\mathbf{b} - D^{-1}L\mathbf{x}^{(r+1)} - (I + D^{-1}U)\mathbf{x}^{(r)}\},$$

or $$\mathbf{x}^{(r+1)} = -\omega D^{-1}L\mathbf{x}^{(r+1)} - [(\omega - 1)I + \omega D^{-1}U]\mathbf{x}^{(r)} + \omega D^{-1}\mathbf{b}. \quad (8)$$

4.3.2 The Basic Convergence Criterion

In equations (6), (7) and (8) we have matrix representations of the Jacobi, Gauss-Seidel and S.O.R. processes respectively. If these relations are written out in full, the results are the systems of recurrence relations which are actually used in practice in the application of the three methods. For the purpose of analysis, however, it is more convenient to have the equations in a form which gives $\mathbf{x}^{(r+1)}$ explicitly in terms of $\mathbf{x}^{(r)}$. Equation (6) is already in this form, and equations (7) and (8) may be expressed similarly on being premultiplied by D and then solved for $\mathbf{x}^{(r+1)}$. The three methods are then represented by

Jacobi: $$\mathbf{x}^{(r+1)} = -D^{-1}(L + U)\mathbf{x}^{(r)} + D^{-1}\mathbf{b}, \quad (9)$$

Gauss-Seidel: $$\mathbf{x}^{(r+1)} = -(D + L)^{-1}U\mathbf{x}^{(r)} + (D + L)^{-1}\mathbf{b}, \quad (10)$$

S.O.R.: $$\mathbf{x}^{(r+1)} = -(D + \omega L)^{-1}[(\omega - 1)D + \omega U]\mathbf{x}^{(r)} + \omega(D + \omega L)^{-1}\mathbf{b}. \quad (11)$$

These equations have a common feature in that they are all of the form

$$\mathbf{x}^{(r+1)} = M\mathbf{x}^{(r)} + \mathbf{c},$$

where M is an $n \times n$ matrix and $\mathbf{c}$ is an $n \times 1$ column vector. In all three cases M and $\mathbf{c}$ are independent of r, and the processes are said to be *stationary*, in the sense that the iterative relations used do not vary from one iteration to the next. It is not difficult to envisage non-stationary processes; for instance, equation (11) would become a non-stationary formula if ω, the over-relaxation parameter, were made to depend upon r, the number of iterations performed. It is worth noting that the method of residual correction, mentioned in the last chapter as a means of improving approximate solutions of linear equations, is also a stationary iterative process, satisfying

$$\mathbf{x}^{(r+1)} = \mathbf{x}^{(r)} + B^{-1}(\mathbf{b} - A\mathbf{x}^{(r)}),$$

or $$\mathbf{x}^{(r+1)} = (I - B^{-1}A)\mathbf{x}^{(r)} + B^{-1}\mathbf{b}, \quad (12)$$

where B^{-1} is an approximation to A^{-1}.

For convergence of an iterative process

$$\mathbf{x}^{(r+1)} = M\mathbf{x}^{(r)} + \mathbf{c}, \quad (13)$$

we require that

$$\lim_{r\to\infty} \mathbf{x}^{(r+1)} = \lim_{r\to\infty} \mathbf{x}^{(r)} = \mathbf{x} = A^{-1}\mathbf{b},$$

and hence the *fixed point*, $\mathbf{x}$, of the process must satisfy

$$\mathbf{x} = M\mathbf{x} + \mathbf{c}. \tag{14}$$

Let the error in the rth iterated solution be given by

$$\boldsymbol{\epsilon}^{(r)} = \mathbf{x}^{(r)} - \mathbf{x}.$$

Subtraction of equation (14) from equation (13) then gives $\boldsymbol{\epsilon}^{(r+1)} = M\boldsymbol{\epsilon}^{(r)}$, whence

$$\boldsymbol{\epsilon}^{(r)} = M^r \boldsymbol{\epsilon}^{(0)}, \tag{15}$$

where $\boldsymbol{\epsilon}^{(0)}$, being the error in an arbitrary initial approximation $\mathbf{x}^{(0)}$, is itself arbitrary. In practice, it is usual to take $\mathbf{x}^{(0)} = \mathbf{0}$ unless better information is available. For convergence, we require

$$\lim_{r\to\infty} \boldsymbol{\epsilon}^{(r)} = \mathbf{0},$$

which, since $\boldsymbol{\epsilon}^{(0)}$ in equation (15) is arbitrary, can only be true if

$$\lim_{r\to\infty} M^r = O \qquad \text{(the null matrix).}$$

As is proved in the Appendix, a necessary and sufficient condition for this to be so is that all the eigenvalues of M are less than unity in absolute magnitude. This is easily demonstrated when all the eigenvectors of M are linearly independent, in which case they can be assembled as a non-singular matrix Q such that $Q^{-1}MQ = D = \text{diag}\,(\lambda_1, \lambda_2, \ldots, \lambda_n)$,* the diagonal matrix of eigenvalues of M. Then also $M = QDQ^{-1}$, whence

$$M^r = QDQ^{-1}.QDQ^{-1}.\ \ldots\ .QDQ^{-1} = QD^rQ^{-1}.$$

Now $D^r = \text{diag}\,(\lambda_1^r, \lambda_2^r, \ldots, \lambda_n^r)$, and hence both $D^r \to O$ and $M^r \to O$ as $r \to \infty$ provided $|\lambda_i| < 1, i = 1, 2, \ldots, n$.

Since $\max_i |\lambda_i|$ is referred to as the *spectral radius* of M, and denoted by $\rho(M)$, we may summarise as follows:

A necessary and sufficient condition for convergence of an iterative process

$$\mathbf{x}^{(r+1)} = M\mathbf{x}^{(r)} + \mathbf{c}$$

is that $\rho(M) < 1$. *A matrix M with this property is said to be* ***convergent.***

Furthermore, convergence will be rapid if the elements of $\boldsymbol{\epsilon}^{(r)}$ tend quickly to zero as r increases. This requires that the elements of M^r and consequently D^r also decrease quickly. But the largest element of D^r has magnitude $\{\rho(M)\}^r$, and it follows that the smaller is $\rho(M)$, the more rapidly will the process ultimately converge.

* Note that in this context D does not denote the diagonal elements of A, as it does elsewhere in this chapter.

4.3.3 Application of the Theory

Returning to the example of Section 4.2.1, we have

$$A = \begin{pmatrix} 2 & -1 & 0 & 0 \\ -1 & 2 & -1 & 0 \\ 0 & -1 & 2 & -1 \\ 0 & 0 & -1 & 2 \end{pmatrix}, \tag{16}$$

and we first express this in the form $A = L + D + U$, where

$$L = \begin{pmatrix} 0 & 0 & 0 & 0 \\ -1 & 0 & 0 & 0 \\ 0 & -1 & 0 & 0 \\ 0 & 0 & -1 & 0 \end{pmatrix}, \qquad U = \begin{pmatrix} 0 & -1 & 0 & 0 \\ 0 & 0 & -1 & 0 \\ 0 & 0 & 0 & -1 \\ 0 & 0 & 0 & 0 \end{pmatrix},$$

and $D = \text{diag}\,(2, 2, 2, 2)$.

The Jacobi process is

$$\mathbf{x}^{(r+1)} = M_J\mathbf{x}^{(r)} + \mathbf{c},$$

where

$$M_J = -D^{-1}(L + U) = \begin{pmatrix} 0 & \frac{1}{2} & 0 & 0 \\ \frac{1}{2} & 0 & \frac{1}{2} & 0 \\ 0 & \frac{1}{2} & 0 & \frac{1}{2} \\ 0 & 0 & \frac{1}{2} & 0 \end{pmatrix}. \tag{17}$$

The characteristic equation of this matrix is

$$16\lambda^4 - 12\lambda^2 + 1 = 0,$$

from which the eigenvalues are found to be ± 0.3090, ± 0.8090. The spectral radius of the Jacobi iteration matrix M_J is therefore $\rho_J = 0.8090$; the fact that ρ_J is only slightly less than unity explains the slow convergence encountered in Section 4.2.1.

For the Gauss-Seidel process we have

$$M_G = -(D + L)^{-1}U = \begin{pmatrix} 0 & \frac{1}{2} & 0 & 0 \\ 0 & \frac{1}{4} & \frac{1}{4} & 0 \\ 0 & \frac{1}{8} & \frac{1}{8} & \frac{1}{2} \\ 0 & \frac{1}{16} & \frac{1}{8} & \frac{1}{4} \end{pmatrix}, \tag{18}$$

with characteristic equation

$$16\lambda^4 - 12\lambda^3 + \lambda^2 = 0.$$

The eigenvalues are now $\lambda = 0, 0, 0.0955, 0.6545$. Thus the spectral radius of M_G is $\rho_G = 0.6545$, and we may expect the Gauss-Seidel process to converge rather more rapidly than the Jacobi. Comparison of the characteristic equations of M_J and M_G shows that, in fact, $\rho_G = \rho_J^2$; one iteration of Gauss-Seidel, in the present instance, is roughly equivalent to two iterations of Jacobi, as we found in practice in Section 4.2.2.

The iteration matrix for S.O.R. is

$$M_s(\omega) = -(D + \omega L)^{-1} [(\omega - 1)D + \omega U]$$

$$= \begin{pmatrix} 1-\omega & \frac{1}{2}\omega & 0 & 0 \\ \frac{1}{2}\omega - \frac{1}{2}\omega^2 & 1 - \omega + \frac{1}{4}\omega^2 & \frac{1}{2}\omega & 0 \\ \frac{1}{4}\omega^2 - \frac{1}{4}\omega^3 & \frac{1}{2}\omega - \frac{1}{2}\omega^2 + \frac{1}{8}\omega^3 & 1 - \omega + \frac{1}{4}\omega^2 & \frac{1}{2}\omega \\ \frac{1}{8}\omega^3 - \frac{1}{8}\omega^4 & \frac{1}{4}\omega^2 - \frac{1}{4}\omega^3 + \frac{1}{16}\omega^4 & \frac{1}{2}\omega - \frac{1}{2}\omega^2 + \frac{1}{8}\omega^3 & 1 - \omega + \frac{1}{4}\omega^2 \end{pmatrix},$$

and the characteristic equation is found to be

$$16(\omega - 1 + \lambda)^4 - 12\omega^2\lambda(\omega - 1 + \lambda)^2 + \omega^4\lambda^2 = 0.$$

Note that $\lambda = 0$ is not a solution unless $\omega = 1$, corresponding to the Gauss-Seidel process already examined. For $\lambda \neq 0$ we can divide the equation by $\omega^4\lambda^2$ to get

$$16\left[\frac{(\omega - 1 + \lambda)^2}{\omega^2\lambda}\right]^2 - 12\left[\frac{(\omega - 1 + \lambda)^2}{\omega^2\lambda}\right] + 1 = 0,$$

whence

$$\frac{(\omega - 1 + \lambda)^2}{\omega^2\lambda} = \mu^2, \tag{19}$$

where $\mu = \pm 0.3090$ or ± 0.8090 is one of the eigenvalues of the Jacobi matrix M_J. From this last equation we find

$$\lambda = \tfrac{1}{2}\mu^2\omega^2 - (\omega - 1) \pm \mu\omega\{\tfrac{1}{4}\mu^2\omega^2 - (\omega - 1)\}^{\frac{1}{2}}. \tag{20}$$

With $\omega = 1.2$, as used in Section 4.2.3, equation (20) gives $\lambda = 0.4545, 0.0880, -0.1312 \pm 0.1509i$. The two real roots correspond to the larger value of μ^2. Since the modulus of the complex roots is only 0.2, the spectral radius of $M_s(1.2)$ is $\rho_s(1.2) = 0.4545$.

Clearly, the best possible rate of convergence will occur for that value of ω which minimises $\rho_s(\omega)$. If we rewrite equation (19) as

$$\frac{\omega - 1 + \lambda}{\omega} = \mu\lambda^{\frac{1}{2}}$$

and define

$$u_\omega(\lambda) = \frac{\omega - 1 + \lambda}{\omega} \quad \text{and} \quad v(\lambda) = \mu\lambda^{\frac{1}{2}},$$

then the eigenvalues of $M_S(\omega)$ are given by the points of intersection of the curves $y = u_\omega(\lambda)$ and $y = v(\lambda)$. The former is a straight line passing through the point (1, 1) and whose slope ω^{-1} decreases monotonely with increasing positive ω; the latter represents two parabolae, one for each value of $|\mu|$:

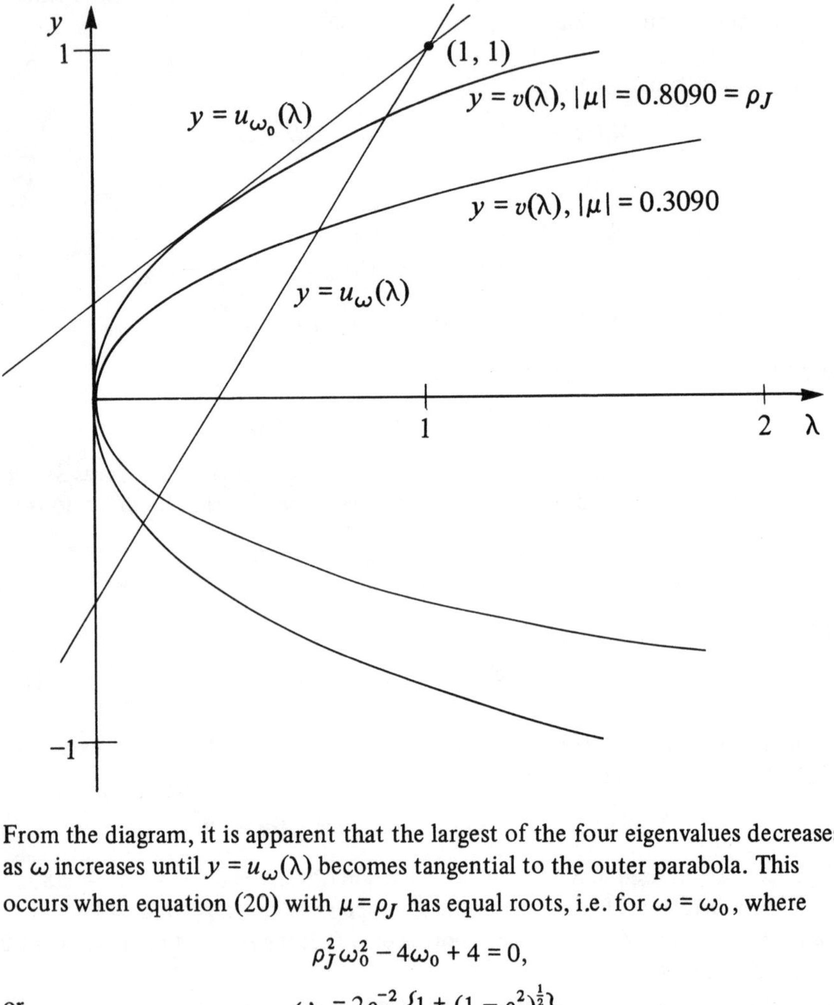

From the diagram, it is apparent that the largest of the four eigenvalues decreases as ω increases until $y = u_\omega(\lambda)$ becomes tangential to the outer parabola. This occurs when equation (20) with $\mu = \rho_J$ has equal roots, i.e. for $\omega = \omega_0$, where

$$\rho_J^2 \omega_0^2 - 4\omega_0 + 4 = 0,$$

or

$$\omega_0 = 2\rho_J^{-2} \{1 \pm (1 - \rho_J^2)^{\frac{1}{2}}\}.$$

We take the negative sign, which gives the smaller value of ω_0; the positive sign gives the other possible point of tangency for which $\lambda > 1$ (a non-convergent case).

For $\omega > \omega_0$ the values of λ become complex; equation (20) gives them as

$$\lambda = \tfrac{1}{2}\mu^2\omega^2 - (\omega - 1) \pm i\mu\omega\{(\omega - 1) - \tfrac{1}{4}\mu^2\omega^2\}^{\frac{1}{2}}.$$

It follows that $|\lambda| = \omega - 1$ for $\omega > \omega_0$ and that S.O.R. will not converge for $\omega \geqslant 2$. Since $|\lambda|$ now increases with increasing ω, whereas for $\omega < \omega_0$ it decreased, we conclude that

$$\omega_0 = 2\rho_J^{-2}\,\{1 - (1 - \rho_J^2)^{\frac{1}{2}}\} = \frac{2}{1 + (1 - \rho_J^2)^{\frac{1}{2}}} \tag{21}$$

minimises the spectral radius of $M_s(\omega)$ and hence gives the best ultimate rate of convergence for S.O.R. Putting $\omega = \omega_0$, $\mu = \rho_J$ and $\lambda = \rho_s(\omega_0)$ in equation (20), we find, after some manipulation, that

$$\rho_s(\omega_0) = \frac{1 - (1 - \rho_J^2)^{\frac{1}{2}}}{1 + (1 - \rho_J^2)^{\frac{1}{2}}} = \omega_0 - 1. \tag{22}$$

Using the value $\rho_J = 0.8090$ in equations (21) and (22) we get

$$\omega_0 = 1.2596 \qquad \text{and} \qquad \rho_s(\omega_0) = 0.2596.$$

Thus a comparatively small change in ω (from 1.2 to 1.2596) produces a very significant reduction in ρ_s (from 0.4545 to 0.2596). With $\omega = \omega_0$, the S.O.R. process applied to equations (1) gives the following:

r	$x_1^{(r)}$	$x_2^{(r)}$	$x_3^{(r)}$	$x_4^{(r)}$	r	$x_1^{(r)}$	$x_2^{(r)}$	$x_3^{(r)}$	$x_4^{(r)}$
0	0.00	0.00	0.00	0.00	5	3.85	6.86	7.93	5.97
1	0.63	1.66	2.93	4.37	6	3.95	6.96	7.98	5.99
2	1.51	3.63	6.16	5.27	7	3.99	6.99	7.99	6.00
3	2.52	5.79	7.25	5.72	8	4.00	7.00	8.00	6.00
4	3.62	6.60	7.77	5.93	9	4.00	7.00	8.00	6.00

Only 8 iterations are needed for two decimal place accuracy.

Let us summarise the principal findings of this section. In the following short table, the second column gives the spectral radius of the appropriate iteration matrix in terms of ρ_J, while the third column gives its numerical value for the present system of equations:

Jacobi	ρ_J	0.8090
Gauss-Seidel	ρ_J^2	0.6545
S.O.R. with $\omega = \omega_0$	$\dfrac{1 - (1 - \rho_J^2)^{\frac{1}{2}}}{1 + (1 - \rho_J^2)^{\frac{1}{2}}}$	0.2596

Here the second column is of interest because these relationships between the spectral radii of the three iteration matrices hold, not just for the present example, but for a large class of similar problems having two characteristics which we discuss in what follows, Property (A) and consistent ordering.

4.3.4 S.O.R. with a Block-tridiagonal System

In practice, particularly in connection with the numerical solution of ordinary and partial differential equations, systems of equations may arise whose coefficient matrix can be partitioned in the *block-tridiagonal* form

$$A = \begin{pmatrix} D_1 & F_1 & & & \\ E_2 & D_2 & F_2 & & \\ & E_3 & D_3 & F_3 & \\ & & \cdots & \cdots & \cdots \\ & & & E_n & D_n \end{pmatrix}, \tag{23}$$

in which the D's are non-singular diagonal matrices, not necessarily all of the same order, while the E's and F's are rectangular matrices. The following important result holds for such matrices:

Theorem

If $A = L + D + U$ has the form of equation (23), then the matrix $-D^{-1}(\alpha L + \alpha^{-1} U)$, derived from the Jacobi iteration matrix $-D^{-1}(L + U)$, has eigenvalues which are independent of the non-zero scalar parameter.

Proof

The Jacobi iteration matrix of A is

$$M_J = -\begin{pmatrix} 0 & D_1^{-1}F_1 & & & \\ D_2^{-1}E_2 & 0 & D_2^{-1}F_2 & & \\ & D_3^{-1}E_3 & 0 & D_3^{-1}F_3 & \\ & & \cdots & \cdots & \cdots \\ & & & D_n^{-1}E_n & 0 \end{pmatrix}.$$

A suitable partitioning of the vector $\mathbf{x}$ in the form $(X_1, X_2, \ldots, X_n)'$ enables the eigenvalue relation $M_J\mathbf{x} = \lambda\mathbf{x}$ to be written as

$$-D_i^{-1}\{E_i X_{i-1} + F_i X_{i+1}\} = \lambda X_i, \qquad i = 1, 2, \ldots, n,$$

where E_0 and F_{n+1} are taken to be null matrices. If we now write

$$X_j = \alpha^{-j-1} Z_j, \qquad j = 1, 2, \ldots, n,$$

then the former equation becomes

$$-D_i^{-1}\left\{\alpha^{-i} E_i Z_{i-1} + \alpha^{-i-2} F_i Z_{i+1}\right\} = \lambda \alpha^{-i-1} Z_i$$

or, on multiplying by $\alpha^{i+1} \neq 0$,

$$-D_i^{-1}\left\{\alpha E_i Z_{i-1} + \alpha^{-1} F_i Z_{i+1}\right\} = \lambda Z_i.$$

This is equivalent to the matrix equation

$$-\begin{pmatrix} 0 & \alpha^{-1}D_1^{-1}F_1 & & & \\ \alpha D_2^{-1}E_2 & 0 & \alpha^{-1}D_2^{-1}F_2 & & \\ & \alpha D_3^{-1}E_3 & 0 & \alpha^{-1}D_3^{-1}F_3 & \\ & \cdots & \cdots & \cdots & \cdots \\ & & & \alpha D_n^{-1}E_n & 0 \end{pmatrix} \begin{pmatrix} Z_1 \\ Z_2 \\ Z_3 \\ \vdots \\ Z_n \end{pmatrix} = \lambda \begin{pmatrix} Z_1 \\ Z_2 \\ Z_3 \\ \vdots \\ Z_n \end{pmatrix},$$

where the left-hand matrix is the modified Jacobi matrix of the theorem. We conclude that this matrix has the same eigenvalues λ as M_J itself, i.e. that the eigenvalues are independent of α.

If $A = L + D + U$ then the corresponding S.O.R. iteration matrix

$$M_s(\omega) = -(D + \omega L)^{-1}\ [(\omega - 1)D + \omega U]$$

has the characteristic equation

$$\det\{-(D + \omega L)^{-1}\ [(\omega - 1)D + \omega U] - \lambda I\} = 0.$$

Equivalently, since $(D + \omega L)$ is non-singular,

$$\det\{-(\omega - 1)D - \omega U - \lambda(D + \omega L)\} = 0, \qquad (24)$$

whence, provided $\lambda \neq 0$,

$$\det\{-\omega\lambda^{\frac{1}{2}}\{\lambda^{\frac{1}{2}}L + \lambda^{-\frac{1}{2}}U\} - (\omega - 1 + \lambda)D\} = 0$$

and

$$\det\left\{-D^{-1}\{\lambda^{\frac{1}{2}}L + \lambda^{-\frac{1}{2}}U\} - \frac{\omega - 1 + \lambda}{\omega\lambda^{\frac{1}{2}}}\, I\right\} = 0.$$

It follows that, if λ is any non-zero eigenvalue of $M_s(\omega)$, then $\mu = (\omega - 1 + \lambda)/\omega\lambda^{\frac{1}{2}}$ is an eigenvalue of $-D^{-1}\ (\lambda^{\frac{1}{2}}L + \lambda^{-\frac{1}{2}}U)$. However, if A is of the form of equation (23), we have established that the eigenvalues of the latter matrix are identical with those of the Jacobi matrix $-D^{-1}(L + U)$. The S.O.R. eigenvalues λ and the Jacobi eigenvalues μ are therefore connected by the relation

$$\mu^2 = \frac{(\omega - 1 + \lambda)^2}{\omega^2\lambda}. \qquad (25)$$

This is identical with equation (19) of Section 4.3.3, although derived under much more general conditions. If all eigenvalues μ of M_J are real, as when A is symmetric, the analysis given there may now be followed exactly. We conclude that, for *any* symmetric matrix of the form of equation (23),

$$\omega = \omega_0 = \frac{2}{1 + (1 - \rho_J^2)^{\frac{1}{2}}}$$

gives $\rho_s(\omega)$ its minimum value of $\omega_0 - 1$, and hence results in the most rapid convergence of S.O.R. If M_J has any complex eigenvalues, the theory is rather more complicated.

The above analysis is valid for $\lambda \neq 0$. However, from equation (24), $\lambda = 0$ can only be a solution if the upper triangular matrix $\{(\omega - 1)D + \omega U\}$ is singular, which will be so only if one or more of its diagonal elements is zero. But we have assumed D to have no zeros on its diagonal, and therefore $\omega = 1$ is the sole value for which $M_S(\omega)$ has zero eigenvalues. This, of course, is the Gauss-Seidel case. The non-zero eigenvalues of M_G, given by equation (25) with $\omega = 1$, are $\lambda = \mu^2$; it follows that $\rho_G = \rho_J^2$ (see, for example, the case of the matrix A of equation (16)).

4.3.5 Property (A) and Consistent Ordering

The results of the foregoing theory carry over to a more general class of matrices characterised by two properties known as *Property (A)* and *consistent ordering.* In order to discuss these we need the concept of a *permutation matrix.* This is a matrix, P, obtained by permuting the rows of an appropriately sized unit matrix I. Some useful properties of permutation matrices are as follows:

(i) P' is the corresponding permutation of the *columns of* I,
(ii) PA and AP' are, respectively, the corresponding permutations of the rows and the columns of A,
(iii) $P^{-1} = P'$, i.e. P is always an orthogonal matrix.

Definition

A matrix A has Property (A) if there exists a permutation matrix P such that $PAP' = T$, where T has the block-tridiagonal form of equation (23), with diagonal submatrices of diagonal form. T will be referred to as a ***tridiagonal representation*** *of A.*

In the context of a system of equations, such a transformation may be effected by first writing the equations in a different order and then reordering the variables correspondingly within the equations.

A system $$A\mathbf{x} = \mathbf{b}$$

becomes $$PA\mathbf{x} = P\mathbf{b}$$

on premultiplication by P. Since $P'P = I$, we can write this as

$$PAP'.P\mathbf{x} = T\mathbf{x}^* = \mathbf{b}^*,$$

where $\mathbf{x}^* = P\mathbf{x}$ and $\mathbf{b}^* = P\mathbf{b}$ are corresponding permutations of the elements of $\mathbf{x}$ and $\mathbf{b}$. The solutions of the original and reordered systems therefore differ only in the order of the components within the vectors.

Ordinarily, if A has Property (A) then there are many suitable matrices P, each giving rise to a different tridiagonal representation. In particular, a representation of the form

$$\begin{pmatrix} D_1 & E \\ F & D_2 \end{pmatrix},$$

in which D_1 and D_2 are square diagonal sub-matrices, is always possible. This fact is sometimes used as an alternative characterisation of Property (A).

Definition

The Property (A) matrix A is said to be ***consistently ordered with respect to*** *its tridiagonal representation T if the actual arithmetic involved in the iterative solution of $A\mathbf{x} = \mathbf{b}$ and $T\mathbf{x}^* = \mathbf{b}^*$ is identical in every respect except for the sequence in which the individual operations are performed.*

In particular, this implies that the rth iterated solutions of the two systems satisfy

$$\mathbf{x}^{*(r)} = P\mathbf{x}^{(r)}$$

for all $r > 0$, provided we start with $\mathbf{x}^{*(0)} = P\mathbf{x}^{(0)}$. Then we must also require

$$\boldsymbol{\epsilon}^{*(r)} = P\boldsymbol{\epsilon}^{(r)},$$

the error vectors being defined, as before, by $\boldsymbol{\epsilon}^{(r)} = \mathbf{x}^{(r)} - \mathbf{x}$ and $\boldsymbol{\epsilon}^{*(r)} = \mathbf{x}^{*(r)} - \mathbf{x}^*$. From the theory of Section 4.3.2, successive error vectors satisfy

$$\boldsymbol{\epsilon}^{(r+1)} = M(A)\boldsymbol{\epsilon}^{(r)}$$

and

$$\boldsymbol{\epsilon}^{*(r+1)} = M(T)\boldsymbol{\epsilon}^{*(r)},$$

where the M's are the relevant iteration matrices. The first of these gives

$$P\boldsymbol{\epsilon}^{(r+1)} = PM(A)\boldsymbol{\epsilon}^{(r)} = PM(A)P'.P\boldsymbol{\epsilon}^{(r)},$$

and comparison with the second then shows that, for arbitrary $\boldsymbol{\epsilon}^{(0)}$, our requirement that $\boldsymbol{\epsilon}^{*(r)} = P\boldsymbol{\epsilon}^{(r)}$ for all $r > 0$ holds if, and only if,

$$M(T) = PM(A)P'.$$

This relation is automatically satisfied for the Jacobi method, since

$$\begin{aligned} M_J(T) &= -D_T^{-1}(L_T + U_T) = -D_T^{-1}(T - D_T) = I - D_T^{-1}T \\ &= PP' - PD_A^{-1}P'.PAP' = P(I - D_A^{-1}A)P' = PM_J(A)P'. \end{aligned}$$

Here we have used the fact that multiplication of a diagonal matrix on the left and the right by P and P' respectively preserves the diagonal form, so that $D_T = PD_AP'$ and $D_T^{-1} = PD_A^{-1}P'$.

For the S.O.R. case the condition for identical arithmetic becomes $M_s(T) = PM_s(A)P'$, or (see equation (11)),

$$\begin{aligned} (D_T + \omega L_T)^{-1}[(\omega - 1)D_T + \omega U_T] &= P(D_A + \omega L_A)^{-1}[(\omega - 1)D_A + \omega U_A]P' \\ &= P(D_A + \omega L_A)^{-1}P'.P[(\omega - 1)D_A + \omega U_A]P' \\ &= [P(D_A + \omega L_A)P']^{-1}[(\omega - 1)PD_AP' + \omega PU_AP'] \\ &= (D_T + \omega PL_AP')^{-1}[(\omega - 1)D_T + \omega PU_AP']. \end{aligned}$$

Here we have again used the fact that $D_T = PD_AP'$. There follows

$$(I + \omega D_T^{-1}L_T)^{-1}[(\omega - 1)I + \omega D_T^{-1}U_T] = (I + \omega D_T^{-1}.PL_AP')^{-1}[(\omega - 1)I + \omega D_T^{-1}.PU_AP'],$$

whence

$$(I + \omega D_T^{-1}.PL_AP')(I + \omega D_T^{-1}L_T)^{-1}[(\omega - 1)I + \omega D_T^{-1}U_T] = [(\omega - 1)I + \omega D_T^{-1}.PU_AP']. \quad (26)$$

On the left-hand side we may expand $(I + \omega D_T^{-1}L_T)^{-1}$ in a power series in $\omega D_T^{-1}L_T$. Since L_T is a strictly lower triangular matrix, $D_T^{-1}L_T$ also has that form, and a characteristic of such matrices is that, if their order is $n \times n$, their n^{th} and higher powers are the $n \times n$ null matrix. Hence the series referred to above terminates, and no questions of convergence arise. We find

$$(I + \omega D_T^{-1}.PL_AP')\{I - \omega(D_T^{-1}L_T) + \omega^2(D_T^{-1}L_T)^2 - \ldots + (-\omega)^{n-1}(D_T^{-1}L_T)^{n-1}\} \times [\omega(I + D_T^{-1}U_T) - I] = [\omega(I + D_T^{-1}.PU_AP') - I],$$

and since we require this to hold for a range of values of ω, we may equate coefficients of powers of ω. From the first-order terms we find that $M_J(T) = PM_J(A)P'$ is a necessary condition for the validity of equation (26), and we have already shown that this is true. The second-order terms give

$$-D_T^{-1}L_T(I + D_T^{-1}U_T) - (D_T^{-1}L_T)^2 + D_T^{-1}.PL_AP'[I + D_T^{-1}U_T + D_T^{-1}L_T] = O,$$

whence
$$(PL_AP' - L_T)D_T^{-1}(L_T + D_T + U_T) = O,$$

Now $L_T + D_T + U_T = T$, which is by hypothesis non-singular. Hence postmultiplication by $T^{-1}D_T$ shows that

$$L_T = PL_AP'$$

is a necessary condition for equation (26) to hold. Substitution for L_T in that equation then shows that it is satisfied for all ω if, and only if,

$$U_T = PU_AP'.$$

These two conditions, then, are necessary and sufficient for S.O.R. to give rise to identical arithmetic when used with either the system $A\mathbf{x} = \mathbf{b}$ or its reordered version $T\mathbf{x}^* = \mathbf{b}^*$. In other words, these are the conditions which must hold for A to be consistently ordered with respect to T.

The significance of the conditions derived above is not difficult to interpret. In the $(r + 1)^{\text{th}}$ iteration of the Gauss-Seidel or S.O.R. solution of $A\mathbf{x} = \mathbf{b}$, the newly calculated components of $\mathbf{x}^{(r+1)}$ are associated with the elements of L_A, while the components of $\mathbf{x}^{(r)}$ remaining from the previous iteration are associated with the elements of U_A (see Section 4.3.1). Clearly, if the transformation $PAP' = T$ transfers a non-zero element from L_A into U_T, this element will now be associated with a component of $\mathbf{x}^{(r+1)}$ in the original system but a component of $\mathbf{x}^{*(r)}$ in the reordered system. Identical arithmetic is therefore not possible unless the only elements of L_A transferred into U_T are zeros. All the non-zero elements of L_A must therefore end up in L_T, as must those of U_A end up in U_T. In such a case, we must have $L_T = PL_AP'$ and $U_T = PU_AP'$, as shown above.

We now generalise the concept of consistent ordering, as follows:

Definition

A Property (A) matrix is said simply to be **consistently ordered** *if it is consistently ordered with respect to at least one of its tridiagonal representations.*

It is quite possible, in fact, for a Property (A) matrix to have many different tridiagonal representations, but to be consistently ordered with respect to none of them (see Section 4.5 for an example). For a consistently ordered Property (A) matrix, then, there exists at least one permutation matrix P such that, not only does $PAP' = T$, but that also A is consistently ordered with respect to T. It follows that

$$\begin{aligned} -D_T^{-1}(\alpha L_T + \alpha^{-1} U_T) &= -PD_A^{-1}P'(\alpha PL_AP' + \alpha^{-1}PU_AP') \\ &= P[-D_A^{-1}(\alpha L_A + \alpha^{-1}U_A)]P', \end{aligned}$$

since the consistent ordering enables us to write $L_T = PL_AP'$ and $U_T = PU_AP'$. The first of these matrices has eigenvalues which are independent of α by virtue of the form of T (as we showed in Section 4.3.4). The third matrix is similar to, and hence has the same eigenvalues as, the correspondingly modified Jacobi matrix of A, $-D_A^{-1}(\alpha L_A + \alpha^{-1}U_A)$. The eigenvalues of this matrix are therefore also independent of α, despite the fact that A is, in general, not of the form of equation (23). An alternative definition of consistent ordering is thus as follows:

Definition

A matrix A is consistently ordered if its modified Jacobi matrix $-D_A^{-1}(\alpha L_A + \alpha^{-1}U_A)$ has eigenvalues independent of the non-zero scalar α.

But this was precisely the property of the matrix of equation (23) which we used earlier to find a relation between the spectral radii of its Jacobi and S.O.R. iteration matrices. It follows that the same relation exists between the spectral radii of those matrices for any consistently ordered matrix with Property (A).

To summarise, we have found that any symmetric matrix with Property (A) and which is consistently ordered has Jacobi, Gauss-Seidel and optimum S.O.R.

iteration matrices whose spectral radii are related by

$$\left.\begin{aligned} \rho_G &= \rho_J^2 \\ \text{and} \qquad \rho_s(\omega_0) &= \frac{1-(1-\rho_J^2)^{\frac{1}{2}}}{1+(1-\rho_J^2)^{\frac{1}{2}}} = \omega_0 - 1, \\ \text{where} \qquad \omega_0 &= \frac{2}{1+(1-\rho_J^2)^{\frac{1}{2}}}\,. \end{aligned}\right\} \tag{27}$$

For other values of ω, $\rho_S(\omega)$ is given in terms of $\mu = \rho_J$ by equation (25). If $\rho_J < 1$ in these relations, it is clear that $\rho_G < \rho_J$. Further,

$$\rho_s(\omega_0) = \frac{1-(1-\rho_J^2)^{\frac{1}{2}}}{1+(1-\rho_J^2)^{\frac{1}{2}}} < 1-(1-\rho_J^2)^{\frac{1}{2}} < 1-(1-\rho_J^2) = \rho_G .$$

Thus

$$\rho_J > \rho_G > \rho_s(\omega_0), \tag{28}$$

and convergence of the Jacobi method ensures convergence of the other two, S.O.R. having the best ultimate rate of convergence.

4.4 Practical Aspects of Convergence

4.4.1 Preliminary Considerations

When setting out to solve a system of equations $A\mathbf{x} = \mathbf{b}$ by an iterative method, we must first answer two questions:

(i) Which method do we wish to use?
and (ii) Will it converge?

Of the three methods examined here, Jacobi's will usually be rejected as having less rapid convergence than Gauss-Seidel (though this is not invariably so). The choice of S.O.R., if still more rapid convergence is sought, raises the problem of finding the optimum value ω_0 of the relaxation parameter. This is often no mean task, and S.O.R. is therefore best reserved for problems where either ω_0 can be easily calculated or where its determination, though not simple, accounts for only a relatively small part of the total computational effort. This might be the case, for example, when $A\mathbf{x} = \mathbf{b}$ is to be solved for a large number of different vectors $\mathbf{b}$.

As regards convergence, the basic criterion is that of Section 4.3.2; the relevant iteration matrix M must have spectral radius $\rho(M)$ less than unity. Unfortunately, this test is difficult to apply in practice for large matrices, since the determination of the largest eigenvalue of M might well involve more computation than the actual solution of our equations. However, certain tests do exist which are simple to apply, though they usually give imprecise information concerning *rate* of convergence.

4.4.2 Convergence of Jacobi and Gauss-Seidel Iteration

For these two methods, we have the following criteria:

(i) Gerschgorin's theorem (see Chapter 7) and one of its corollaries give upper bounds to the spectral radius of a matrix M. If M is of order $n \times n$, the theorem itself states that the eigenvalues all lie (in the complex z-plane) in the union of the discs

$$|z - m_{ii}| \leqslant S_i, \qquad i = 1, 2, \ldots, n,$$

where

$$S_i \equiv \sum_{\substack{j=1 \\ j \neq i}}^{n} |m_{ij}|$$

is the sum of the moduli of the elements of the ith row of M, excluding the diagonal element. Since the disc $|z - m_{ii}| \leqslant S_i$ is contained in the disc

$$|z| \leqslant |m_{ii}| + S_i = \sum_{j=1}^{n} |m_{ij}|,$$

centred at the origin, we have the less precise but more easily applied corollary

$$\rho(M) \leqslant \max_i \sum_{j=1}^{n} |m_{ij}|.$$

As M and its transpose M' have the same eigenvalues, these results may equally well be applied to column sums, which sometimes give a smaller bound on $\rho(M)$. Consider, for example, the Gauss-Seidel iteration matrix of equation (18). This has row modulus sums $\frac{1}{2}$, $\frac{3}{4}$, $\frac{7}{8}$ and $\frac{7}{16}$, and column modulus sums 0, $\frac{15}{16}$, $\frac{7}{8}$, and $\frac{3}{4}$. In this example, then, the row sums give the better bound on $\rho(M_G)$, namely $\rho(M_G) \leqslant \frac{7}{8}$.

Normally, we would not wish to calculate the Gauss-Seidel iteration matrix, since this involves the inversion of $(D + L)$. The Jacobi matrix is always easy to calculate, however, and in cases where the inequalities (28) hold, $\rho(M_J) < 1$ implies convergence also of the Gauss-Seidel process. For a general $n \times n$ system, $A\mathbf{x} = \mathbf{b}$, the Jacobi matrix is (see equation (9))

$$M_J = \begin{pmatrix} 0 & -\dfrac{a_{12}}{a_{11}} & -\dfrac{a_{13}}{a_{11}} & \cdots & -\dfrac{a_{1n}}{a_{11}} \\ -\dfrac{a_{21}}{a_{22}} & 0 & -\dfrac{a_{23}}{a_{22}} & \cdots & -\dfrac{a_{2n}}{a_{22}} \\ \cdots & \cdots & \cdots & \cdots & \cdots \\ -\dfrac{a_{n1}}{a_{nn}} & -\dfrac{a_{n2}}{a_{nn}} & -\dfrac{a_{n3}}{a_{nn}} & \cdots & 0 \end{pmatrix},$$

and the row sum form of the Gerschgorin corollary gives

$$\rho(M_J) \leqslant \max_i \sum_{\substack{j=1 \\ j \neq i}}^{n} \left| \frac{a_{ij}}{a_{ii}} \right|.$$

Then a sufficient condition for convergence is for the summation to be <1 for all i, i.e.

$$|a_{ii}| > \sum_{\substack{j=1 \\ j \neq i}}^{n} |a_{ij}| \qquad \text{for } i = 1, 2, \ldots, n. \tag{29}$$

A matrix A having this property is *strictly diagonally dominant*, as defined in Section 4.2.4.

The failure of this test does not necessarily imply non-convergence of the process in question; the reader may like to verify, for instance, that the test fails for the Jacobi matrix of equation (17)*, which we know to be convergent. In some cases of failure, it may be possible to rearrange the system of equations to bring large elements of A onto the diagonal, enabling the test to be successfully reapplied.

(ii) If A is a symmetric, positive-definite matrix, then its Gauss-Seidel iteration matrix $M_G = -(D + L)^{-1}U$ has spectral radius less than unity. Since symmetry of A does not imply that M_G is symmetric, the latter may have complex eigenvalues and eigenvectors, say λ and $\mathbf{y}$. These must satisfy

$$-(D + L)^{-1}U\mathbf{y} = \lambda\mathbf{y},$$

or, since $U = L'$ because A is symmetric,

$$L'\mathbf{y} = -\lambda(D + L)\mathbf{y}.$$

Pre-multiplication by $\bar{\mathbf{y}}'$, the transposed conjugate of $\mathbf{y}$, gives

$$\bar{\mathbf{y}}'L'\mathbf{y} = -\lambda\bar{\mathbf{y}}'(D + L)\mathbf{y}. \tag{30}$$

But since M_G is a real matrix $\bar{\lambda}$ is also an eigenvalue, with corresponding eigenvector $\bar{\mathbf{y}}$. Thus, by a similar argument,

$$\mathbf{y}'L'\bar{\mathbf{y}} = -\bar{\lambda}\mathbf{y}'(D + L)\bar{\mathbf{y}}. \tag{31}$$

Now the left-hand side of this relation is a complex number which, regarded as a 1×1 symmetric matrix, is its own transpose. Hence

$$(\mathbf{y}'L'\bar{\mathbf{y}})' = \bar{\mathbf{y}}'L\mathbf{y} = -\bar{\lambda}\mathbf{y}'(D + L)\bar{\mathbf{y}}. \tag{32}$$

Thus, from equations (30) and (32),

$$\left.\begin{aligned} \bar{\mathbf{y}}'L'\mathbf{y} &= p + iq \\ \text{and} \quad \bar{\mathbf{y}}'L\mathbf{y} &= p - iq \end{aligned}\right\}, \tag{33}$$

p and q being real scalars. Adding these relations, and adding a further $\bar{\mathbf{y}}'D\mathbf{y}$ to each side, we obtain

$$\bar{\mathbf{y}}'(L + D + L')\mathbf{y} = \bar{\mathbf{y}}'A\mathbf{y} = 2p + \bar{\mathbf{y}}'D\mathbf{y}, \tag{34}$$

* Varga [4] (p. 73) shows that for matrices of a type called *irreducibly diagonally dominant* the $>$ sign in equation (29) may be replaced by $\geqslant$. By this modified criterion, the Jacobi matrix of equation (17) is convergent.

which is positive since A is positive definite. If we now put equations (33) into equation (30), the latter becomes

$$p + iq = -\lambda\{\bar{\mathbf{y}}'D\mathbf{y} + (p - iq)\},$$

whence
$$\lambda = -\frac{p + iq}{\bar{\mathbf{y}}'D\mathbf{y} + p - iq}$$

and
$$|\lambda|^2 = \frac{p^2 + q^2}{\{\bar{\mathbf{y}}'D\mathbf{y} + p\}^2 + q^2}.$$

For convergence, we require that $|\lambda|^2 < 1$, which will be so if

$$|\bar{\mathbf{y}}'D\mathbf{y} + p| > |p|.$$

Squaring, subtracting p^2 from both sides and cancelling a factor $\bar{\mathbf{y}}'D\mathbf{y}$ (which is positive, since if A is positive definite then so is D), we boil this condition down to

$$\bar{\mathbf{y}}'D\mathbf{y} + 2p > 0.$$

But this is true by virtue of equation (34), and convergence is therefore assured.

In passing, we may note that the matrix of the examples given earlier in this chapter (see equation (16)) is positive-definite, so that the Gauss-Seidel method is guaranteed of success.

(iii) Our last test for convergence is the Stein-Rosenberg comparison theorem (see Varga [4] for the proof), which runs as follows:

If the system $A\mathbf{x} = \mathbf{b}$ *has a Jacobi iteration matrix* $M_J = -D^{-1}(L + U)$ *which contains no negative elements, then either*

(a) $\rho(M_J) = \rho(M_G) = 0,$
(b) $0 < \rho(M_G) < \rho(M_J) < 1,$
(c) $\rho(M_J) = \rho(M_G) = 1,$
or (d) $1 < \rho(M_J) < \rho(M_G).$

The four possibilities are mutually exclusive, and we see that the Jacobi and Gauss-Seidel methods either converge together or diverge together for the type of matrix M_J referred to in the theorem. Furthermore, for a such matrix, if $\rho(M_J) < 1$ the Gauss-Seidel process converges faster than the Jacobi. In illustration, we may consider the Jacobi matrix of equation (17), which contains no negative entries. Since for this matrix we found earlier that $\rho(M_J) = 0.8090$, possibility (b) above applies, confirming that the Gauss-Seidel method has the faster convergence.

4.4.3 Convergence of S.O.R.

The Gauss-Seidel method is a special case of S.O.R. with relaxation parameter $\omega = 1$. Considerations of continuity in ω imply that, if the Gauss-Seidel method converges for a given problem, S.O.R. will converge for a range of values of ω including $\omega = 1$. Somewhere in this range will be the optimum value ω_0 which minimises the spectral radius of the S.O.R. iteration matrix $M_s(\omega)$, and for efficient use of the method we need a means for determining this value. As shown in Section 4.3.5, when a system $A\mathbf{x} = \mathbf{b}$ is such that A has Property (A) and is consistently ordered, ω_0 can be calculated in terms of the Jacobi spectral radius ρ_J. If A has Property (A) but is not consistently ordered, it is always possible to find a permutation matrix P such that PAP' is block-tridiagonal with diagonal submatrices of diagonal form; the S.O.R. theory of Section 4.3.4 then applies. Failing this, we must usually fall back on methods for estimating ω_0 numerically, since the analytical procedure used in Section 4.3.3 for a 4 x 4 system is out of the question for large matrices.

Unfortunately, there is no general means by which a Property (A) matrix may be immediately recognised. The most straightforward method is the following:

(i) Disregard all diagonal elements of the matrix;
(ii) Determine whether the integers $1, 2, \ldots, n$ can be divided into two sets, I and II, such that, for every non-zero off-diagonal element a_{ij}, i and j each fall into different sets.

Consider, for example, the matrix

$$\begin{pmatrix} a_{11} & a_{12} & 0 & a_{14} \\ a_{21} & a_{22} & a_{23} & 0 \\ 0 & a_{32} & a_{33} & a_{34} \\ a_{41} & 0 & a_{43} & a_{44} \end{pmatrix},$$

in which all the a's are non-zero.

From the non-zero elements of the first row (ignoring a_{11}), we see that if 1 is put into set I, then 2 and 4 must go into set II. In the second row, the element a_{21} does not upset this classification, while the element a_{23} dictates that, since 2 is already in set II, 3 must go into set I. We then have

set I	set II
1,3	2,4.

The suffixes of all the remaining non-zero elements of the matrix fall in with this classification. The matrix therefore has Property (A).

Now, is the matrix consistently ordered? This property, like Property (A), is not particularly easy to establish in general. In the present case, if we set up the modified Jacobi matrix $-D^{-1}(\alpha L + \alpha^{-1}U)$ and determine its characteristic equation, we find

$$a_{11}a_{22}a_{33}a_{44}\lambda^4 - (a_{11}a_{23}a_{32}a_{44} + a_{11}a_{22}a_{34}a_{43} + a_{12}a_{21}a_{33}a_{44} + a_{14}a_{22}a_{33}a_{41})\lambda^2$$
$$+ (a_{12}a_{21}a_{34}a_{43} + a_{14}a_{23}a_{32}a_{41} - a_{12}a_{23}a_{34}a_{41}\alpha^{-2} - a_{14}a_{21}a_{32}a_{43}\alpha^2) = 0,$$

whose roots depend upon α. Then the matrix is not consistently ordered, and an initial reordering into a suitable block-tridiagonal form must be made before S.O.R. is applied. One way of doing this makes use of the sets I and II determined in establishing Property (A). If the rows of the matrix are reordered so that all those corresponding to the elements of set I precede all those corresponding to the elements of set II, and the corresponding column reordering is then made, the resulting matrix will have the required form. The reader may like to confirm this for the matrix under consideration.

For large matrices, the direct test used above for consistent ordering is clearly impracticable, and an initial block-tridiagonal reordering may be made to avoid the need for any test. In certain special but very important cases, a simply applied test for consistent ordering does exist, however, and we shall have more to say on this matter in Section 4.5.

For problems involving a matrix of coefficients without Property (A) there is no general theory to help us, though Varga [4] presents some extension of the analysis given earlier to other restricted classes of matrices. One simple but time-consuming procedure for estimating ω_0 is to run the problem on a computer for a range of values of ω to obtain some idea of the value which gives the most rapid convergence. Alternatively, experience has shown that even when the Property (A) theory does not apply, if ρ_J is determined for the matrix in question and substituted into the third of equations (27) as though the theory were still valid, the resulting value of ω is close to the optimum. In practice, it is easier to estimate ρ_G than ρ_J, and we then make the equally invalid assumption that $\rho_J^2 = \rho_G$ in the third of equations (27).

Whether the Property (A) theory applies or not, then, there remains the problem of determining ρ_J or ρ_G. In a small number of rather specialised cases ρ_J can be calculated quite easily; for example, if A is an $n \times n$ tridiagonal matrix whose Jacobi matrix is

$$M_J = \begin{pmatrix} 0 & b & & & \\ a & 0 & b & & \\ & a & 0 & b & \\ & & \cdots & \cdots & \\ & & & a & 0 \end{pmatrix},$$

it is known that $\rho_J = 2(ab)^{\frac{1}{2}} \cos \dfrac{\pi}{(n+1)}$. More usually, some numerical means must be employed for estimating ρ_J or ρ_G. Clearly, this will only be worthwhile if the determination of ω_0 is no more than a small part of some larger computational effort. One simple method for estimating ρ_G is based on the following argument:

Gauss-Seidel iteration is defined by

$$\mathbf{x}^{(r+1)} = M_G \mathbf{x}^{(r)} + \mathbf{c}.$$

It follows that

$$\mathbf{x}^{(r)} = M_G \mathbf{x}^{(r-1)} + \mathbf{c},$$

and if we define the displacement vector $\mathbf{d}^{(r)}$ by $\mathbf{d}^{(r)} = \mathbf{x}^{(r+1)} - \mathbf{x}^{(r)}$, subtraction gives

$$\mathbf{d}^{(r)} = M_G \mathbf{d}^{(r-1)}.$$

It follows that

$$\mathbf{d}^{(r)} = M_G^r \mathbf{d}^{(0)}. \tag{35}$$

Now suppose that M_G is a real $n \times n$ matrix with n linearly independent eigenvectors $\mathbf{v}_i$. Then $\mathbf{d}^{(0)}$ in equation (35) can be written as a linear combination of the $\mathbf{v}_i$, with the result

$$\mathbf{d}^{(r)} = M_G^r \sum_{i=1}^{n} \alpha_i \mathbf{v}_i,$$

where the α_i are scalar constants. It is easily verified that, if λ_j is the eigenvalue corresponding to the eigenvector $\mathbf{v}_j$, then

$$M_G^r \mathbf{v}_j = \lambda_j^r \mathbf{v}_j,$$

whence

$$\mathbf{d}^{(r)} = \sum_{i=1}^{n} \alpha_i \lambda_i^r \mathbf{v}_i.$$

Now assume that M_G has only one eigenvalue of magnitude ρ_G*; if this eigenvalue is λ_1, the above equation shows that, as $r \to \infty$,

$$\mathbf{d}^{(r)} \to \alpha_1 \lambda_1^r \mathbf{v}_1$$

because the first term in the summation becomes dominant as r increases.

Then also $$\mathbf{d}^{(r+1)} \to \alpha_1 \lambda_1^{r+1} \mathbf{v}_1,$$

whence $$\mathbf{d}^{(r+1)} \to \lambda_1 \mathbf{d}^{(r)}. \tag{36}$$

After a large number of iterations, then, we can estimate λ_1 by taking the ratio of corresponding elements of $\mathbf{d}^{(r+1)}$ and $\mathbf{d}^{(r)}$, finally taking $\rho_J^2 = \rho_G = |\lambda_1|$ in equation (27) to get ω_0.

This method for finding the largest eigenvalue of M_G is called the *power method*; it is examined in more detail in Chapter 1, where remedies are suggested for some of the problems which often arise with its use in practice.

If S.O.R. is to be used with a numerically estimated value of ω_0, an important point is that a slightly overestimated value is much to be preferred to a value underestimated by the same amount. For problems with Property (A) which are

* This implies, of course, that λ_1 is real; if it were complex, $\bar{\lambda}_1$, which has the same magnitude, would also be an eigenvalue.

consistently ordered, the relation between $\rho_s(\omega)$ and ω has the form

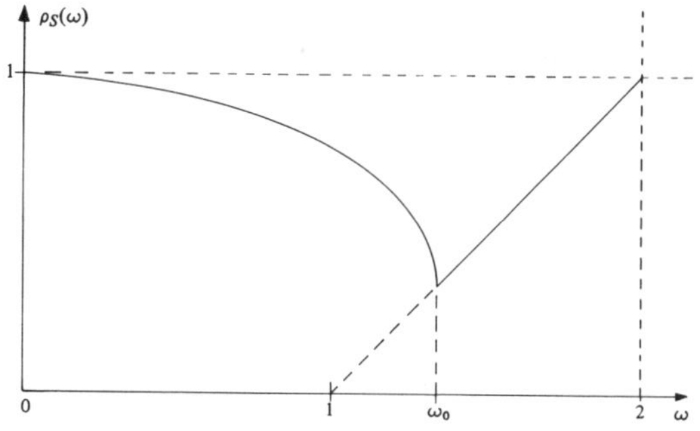

This is established from further consideration of equation (20), and the pertinent feature is that $\rho_s(\omega)$ has an infinite gradient as $\omega \to \omega_0$ from the left but a gradient of only 1 for $\omega > \omega_0$. Experience has shown that this type of variation also generally occurs for problems which do not have Property (A). For small deviations $\delta > 0$, then, $\omega_0 + \delta$ gives a smaller value of the spectral radius than does $\omega_0 - \delta$.

4.4.4 Devices for Accelerating Convergence

It was shown in Section 4.3.2 that successive error vectors for the iterative process $\mathbf{x}^{(r+1)} = M\mathbf{x}^{(r)} + \mathbf{c}$ satisfy $\boldsymbol{\epsilon}^{(r+1)} = M^r\boldsymbol{\epsilon}^{(0)}$. This relation has a strong formal resemblance to equation (35) of Section 4.4.3, and a repetition of the argument advanced there shows that, as r becomes large,

$$\boldsymbol{\epsilon}^{(r+1)} \to \lambda_1 \boldsymbol{\epsilon}^{(r)}. \tag{37}$$

Here it is again assumed that M has distinct eigenvectors, and that λ_1 is the eigenvalue of largest modulus. Whenever equation (37) is valid, simple means are available for accelerating convergence. For instance, we have

$$\mathbf{x}^{(r)} - \mathbf{x} = \boldsymbol{\epsilon}^{(r)} \tag{38}$$

by definition, and for large r we may write

$$\mathbf{x}^{(r+1)} - \mathbf{x} = \boldsymbol{\epsilon}^{(r+1)} \simeq \lambda_1 \boldsymbol{\epsilon}^{(r)}. \tag{39}$$

Elimination of $\boldsymbol{\epsilon}^{(r)}$ between equations (38) and (39) gives

$$\mathbf{x} \simeq \frac{\mathbf{x}^{(r+1)} - \lambda_1 \mathbf{x}^{(r)}}{1 - \lambda_1}, \tag{40}$$

which, if λ_1 is known, should give a better approximation to $\mathbf{x}$ than either $\mathbf{x}^{(r)}$ or $\mathbf{x}^{(r+1)}$. A single application of equation (40) may give a gain in accuracy which would otherwise require several more iterations of the basic process.

If λ_1 is not known, we may re-write equation (37) as

$$\mathbf{x}^{(r+1)} - \mathbf{x} \simeq \lambda_1 (\mathbf{x}^{(r)} - \mathbf{x}).$$

Then for the ith components we have the scalar relation

$$x_i^{(r+1)} - x_i \simeq \lambda_1 (x_i^{(r)} - x_i),$$

whence also

$$x_i^{(r)} - x_i \simeq \lambda_1 (x_i^{(r-1)} - x_i).$$

Elimination of λ_1 from these equations gives

$$x_i \simeq y_i^{(r)} = \frac{x_i^{(r-1)} x_i^{(r+1)} - \{x_i^{(r)}\}^2}{x_i^{(r-1)} - 2x_i^{(r)} + x_i^{(r+1)}}.$$

This manner of combining three successive iterated solutions to generate an improved approximation is known as *Aitken's δ^2-process.*

The use of such acceleration techniques is most appropriate (and, indeed, is sometimes imperative) when convergence is otherwise very slow, as is often the case with problems of large order.

To illustrate the power of the δ^2-process under favourable conditions, we apply it to the results of the Gauss-Seidel solution of Section 4.2.2. For the component x_1, we find the results quoted to six decimal places below:

r	$x_1^{(r)}$	$y_1^{(r)}$
0	0.000000	
1	0.500000	
2	1.125000	
3	1.812500	
4	2.539063	4.162768
5	3.041016	4.013365
6	3.372070	4.001254
7	3.588989	4.000125
8	3.730988	4.000007
9	3.823929	4.000002

We here have the value of x_1 correct to two decimal places after only 6 iterations (compared with the 18 needed without the δ^2-process). More remarkably, a further 3 iterations give x_1 correct to *five* decimal places.

4.4.5 Criteria for Terminating Iterative Processes

Frequently, iterative processes are terminated when some measure of the difference between two successive iterates is found to be less than a specified small number δ. The actual test used may employ either their absolute or their relative difference, as in

$$\max_i |x_i^{(r+1)} - x_i^{(r)}| < \delta?$$

or

$$\max_i \left| \frac{x_i^{(r+1)} - x_i^{(r)}}{x_i^{(r+1)}} \right| < \delta?$$

Unfortunately, these simple criteria can give misleading results, since they do not indicate how well the computed solution actually satisfies the original equations. We saw in the last section that successive error vectors are often related for large r by $\boldsymbol{\epsilon}^{(r+1)} \simeq \lambda_1 \boldsymbol{\epsilon}^{(r)}$, so that if $|\lambda_1|$ is only slightly less than unity the error vector may change almost imperceptibly between iterations. Thus good agreement between $\mathbf{x}^{(r)} = \mathbf{x} + \boldsymbol{\epsilon}^{(r)}$ and $\mathbf{x}^{(r+1)} = \mathbf{x} + \boldsymbol{\epsilon}^{(r+1)}$ is no guarantee that either is close to the true solution $\mathbf{x}$.

By way of example, consider the table on p. 74, which lists the results of a Jacobi process with $\rho_J = 0.8090$, rounded to two decimal places. Note that agreement to that accuracy between successive iterates is achieved in 29 iterations, at which stage two components are still in error by -0.02.

An alternative termination criterion examines the *residual vector* $\boldsymbol{\eta}^{(r)} = \mathbf{b} - A\mathbf{x}^{(r)}$; for instance, the test might be

$$\max_i |\eta_i^{(r)}| < \delta?$$

At first sight this is attractive, in that it tests how well the computed solution fits the equations. However, it is not very satisfactory for ill-conditioned systems. Such systems have the property that a small change in any coefficient results in a very large change in the solution, and conversely that a small change in a computed solution gives rise to a very small change in $\boldsymbol{\eta}$. What is worse, an ill-conditioned system always has $|\lambda_1|$ very close to unity (see Exercise 8 at the end of this chapter), so that convergence is slow and our earlier criteria unreliable. Clearly, such cases must be treated with extreme caution; perhaps the best approach is to keep a check on the residual vector using double-precision arithmetic.

To illustrate the foregoing, we consider an example given by Fox [2] (p. 202):

$$1.00000\, x_1 + 0.70710\, x_2 = 0.29290$$
$$0.70710\, x_1 + 0.50000\, x_2 = 0.20711.$$

The Gauss-Seidel method, using five decimal place arithmetic, gives $x_1 = 0.29290$, $x_2 = 0.00000$ for the first *and all subsequent* iterations. A criterion which tests $|x_i^{(r+1)} - x_i^{(r)}|$ will judge this to be the solution. Further, the residual vector is $\boldsymbol{\eta} = (0.00000000, -0.00000041)'$, both components being zero to 5 decimal places. But in fact the true solution is $\mathbf{x} = (0.26267, 0.04275)'$, the first criterion failing because of slow convergence ($\rho_G = 0.99998$), and the second because of ill-conditioning.

4.5 An Illustrative Example

Many points made earlier in this chapter are illustrated by the following example. We will suppose that we wish to solve Laplace's equation

$$\frac{\partial^2 u}{\partial x^2} + \frac{\partial^2 u}{\partial y^2} = 0 \tag{41}$$

for $u(x, y)$ in the square region $0 < x < 1, 0 < y < 1$, subject to the boundary conditions $u(0, y) = u(1, y) = 0$ for $0 \leqslant y \leqslant 1, u(x, 0) = 0$ and $u(x, 1) = 64(x - x^3)$ for $0 \leqslant x \leqslant 1$. In the most straightforward numerical approach to this problem, the partial derivatives are approximated in terms of function values. For constant y and sufficiently small h we may write the Taylor expansion

$$u(x+h, y) = u(x, y) + hu_x(x, y) + \tfrac{1}{2}h^2 u_{xx}(x, y) + \tfrac{1}{6}h^3 u_{xxx}(x, y) + O(h^4),$$

where a subscript x denotes partial differentiation with respect to x. Similarly,

$$u(x-h, y) = u(x, y) - hu_x(x, y) + \tfrac{1}{2}h^2 u_{xx}(x, y) - \tfrac{1}{6}h^3 u_{xxx}(x, y) + O(h^4).$$

On adding these equations and solving for $u_{xx}(x, y)$, we then have

$$u_{xx}(x, y) = \frac{1}{h^2}\{u(x-h, y) - 2u(x, y) + u(x+h, y)\} + O(h^2).$$

In a similar manner, we obtain

$$u_{yy}(x, y) = \frac{1}{h^2}\{u(x, y-h) - 2u(x, y) + u(x, y+h)\} + O(h^2),$$

and using these two results, we may approximate Laplace's equation, (41), by

$$4u(x, y) - u(x-h, y) - u(x+h, y) - u(x, y-h) - u(x, y+h) = 0. \tag{42}$$

Now let us suppose the region $0 < x < 1, 0 < y < 1$ to be covered by a square mesh, as shown in the following diagram:

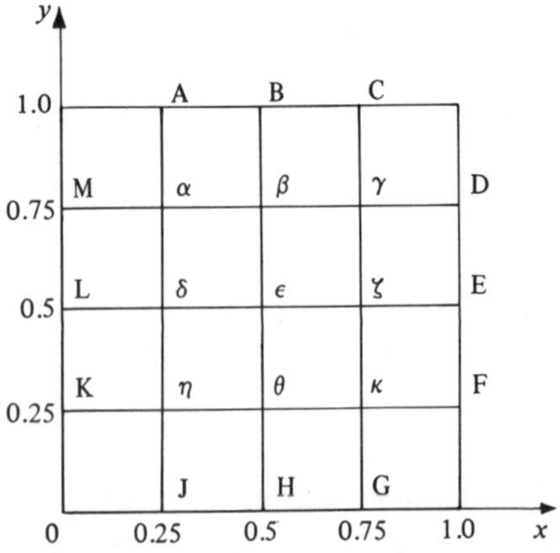

Here the internal mesh intersections have been labelled with the greek characters $\alpha, \beta, \ldots, \kappa$, and the intersections of the mesh with the boundary by the roman A, B, . . . , M. It is clear that equation (42), with $h = 0.25$, is an approximate linear relation between the values of u at five neighbouring mesh intersections; for instance, if (x, y) is the point labelled ϵ, then

$$4u_\epsilon - u_\delta - u_\zeta - u_\theta - u_\beta = 0.$$

From the diagram, we can write nine such relations, one for each internal mesh point. Most of these equations will involve the value of u at one or more of the boundary points A, B, . . . , M. These u-values are specified by the boundary conditions, however, and the only unknowns are the nine internal values $u_\alpha, u_\beta, \ldots, u_\kappa$. We thus have a 9 x 9 system of linear equations to solve.

Let us now relabel the internal mesh points:

	A	B	C	
M	1	2	3	D
L	4	5	6	E
K	7	8	9	F
	J	H	G	

If we write down our system of nine equations, taking the point (x, y) in equation (42) to be the points labelled 1, 2, . . . , 9 successively, the result is

$$\left(\begin{array}{ccc|ccc|ccc}
4 & -1 & & -1 & & & & & \\
-1 & 4 & -1 & & -1 & & & & \\
& -1 & 4 & & & -1 & & & \\
\hline
-1 & & & 4 & -1 & & -1 & & \\
& -1 & & -1 & 4 & -1 & & -1 & \\
& & -1 & & -1 & 4 & & & -1 \\
\hline
& & & -1 & & & 4 & -1 & \\
& & & & -1 & & -1 & 4 & -1 \\
& & & & & -1 & & -1 & 4
\end{array}\right)
\begin{pmatrix} u_1 \\ u_2 \\ u_3 \\ u_4 \\ u_5 \\ u_6 \\ u_7 \\ u_8 \\ u_9 \end{pmatrix}
=
\begin{pmatrix} u_A + u_M \\ u_B \\ u_C + u_D \\ u_L \\ 0 \\ u_E \\ u_J + u_K \\ u_H \\ u_G \end{pmatrix}. \quad (43)$$

The matrix of coefficients can be partitioned in tridiagonal form as shown, though the diagonal submatrices do not have diagonal form, and the S.O.R. theory of Section 4.3.4 does not apply directly. However, the criterion given in Section 4.4.3 shows that the matrix has Property (A), since its non-zero off-diagonal elements a_{ij} all occupy positions such that, if i belongs to the set $\{1, 3, 5, 7, 9\}$, then j belongs to $\{2, 4, 6, 8\}$, and vice versa. As the matrix has

Property (A), a suitable reordering of its rows, followed by the corresponding reordering of its columns, will result in a tridiagonal matrix whose diagonal submatrices are themselves diagonal. In fact, the system may be reorganised in this manner simply by relabelling the internal mesh points. Consider the three following renumberings:

	1	3	6
	2	5	8
	4	7	9

(i)

	1	6	2
	7	3	8
	4	9	5

(ii)

	1	2	3
	6	5	4
	7	8	9

(iii)

If equation (42) is applied to the mesh points in the order in which they are numbered, the resulting matrices have the following patterns of non-zero elements:

(i)

$$\left(\begin{array}{c|cc|ccc|cc|c}
* & * & * & & & & & & \\ \hline
* & * & & * & * & & & & \\
* & & * & & * & * & & & \\ \hline
& * & & * & & & * & & \\
& * & * & & * & & * & * & \\
& & * & & & * & & * & \\ \hline
& & & * & * & & * & & * \\
& & & & * & * & & * & * \\ \hline
& & & & & & * & * & *
\end{array}\right)$$

(ii)

$$\left(\begin{array}{ccccc|cccc}
* & & & & & * & * & & \\
& * & & & & * & & * & \\
& & * & & & * & * & * & * \\
& & & * & & & * & & * \\
& & & & * & & & * & * \\ \hline
* & * & * & & & * & & & \\
* & & * & * & & & * & & \\
& * & * & & * & & & * & \\
& & * & * & * & & & & *
\end{array}\right)$$

(iii)

$$\left(\begin{array}{ccccccccc}
* & * & & & & * & & & \\
* & * & * & & * & & & & \\
& * & * & * & & & & & \\
& & * & * & * & & & & * \\
& * & & * & * & * & & * & \\
* & & & & * & * & * & & \\
& & & & & * & * & * & \\
& & & & * & & * & * & * \\
& & & * & & & & * & *
\end{array}\right)$$

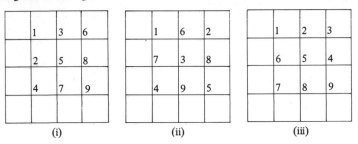

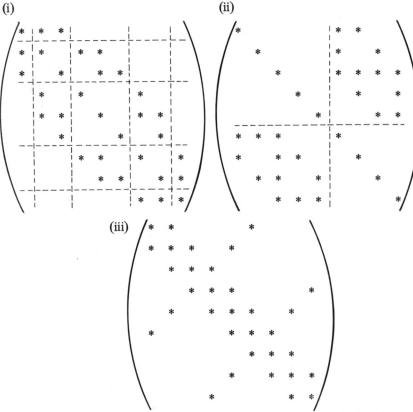

The first two cases give rise to matrices which are, as indicated, block-tridiagonal with diagonal submatrices themselves diagonal. The third case does not permit such a partitioning, but gives rise to a matrix which must possess Property (A), since relabelling the mesh points has the effects of (a) permuting the order of the equations and (b) permuting the variables $u_\alpha, u_\beta, \ldots, u_\kappa$ within the equations in the corresponding manner. In fact, it follows that *any* relabelling of the mesh points in this problem gives rise to a Property (A) matrix.

We turn now to the question of consistent ordering. Young (quoted by Forsythe and Wasow [1] states the following criterion for consistent ordering of equations deriving from a five-point approximation to an elliptic partial differential equation:

If the mesh lines between adjacent intersections are replaced by arrows directed from the smaller label to the larger, and if it is found that each elementary rectangular cell of the mesh is bounded by two arrows corresponding to clockwise and two to anti-clockwise rotation about the interior of the cell, the resulting system of equations is consistently ordered.

As an example, consider case (iii) of the orderings previously examined. The top right-hand interior mesh square gives

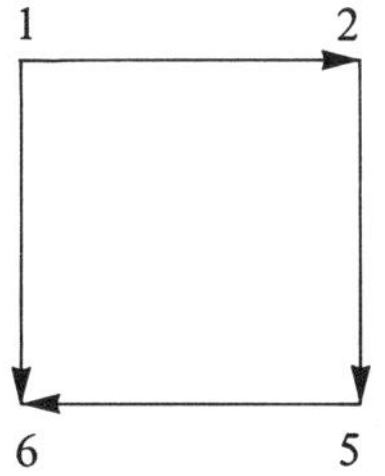

in which three of the arrows point clockwise and one anticlockwise. The ordering of case (iii) is therefore not consistent. On the other hand, the 'natural' mesh labelling leading to equation (43) gives rise to the following arrow pattern:

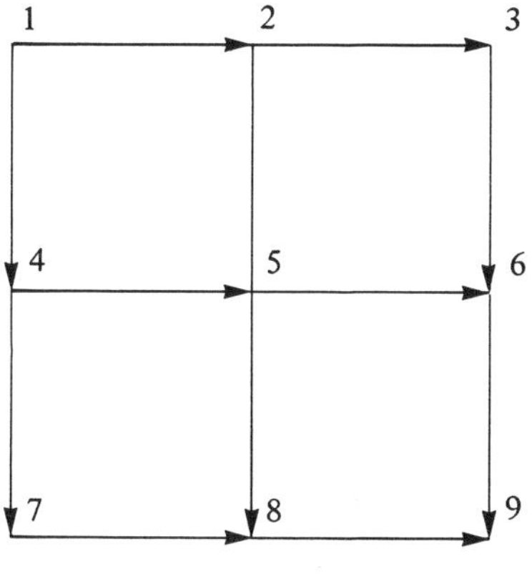

Equation (43), then, is consistently ordered.

As shown in Section 4.3.5, the significance of consistent ordering lies in its guarantee that at least one of the possible reorderings of equation (43) in the block-tridiagonal form of equation (23) has identical iterative properties. In the present context, this implies that, in terms of our earlier, fixed labelling an initially assumed approximate solution $u_\alpha^{(0)}, u_\beta^{(0)}, \ldots, u_\kappa^{(0)}$ gives rise to exactly the same sequence of approximate values for $u_\alpha, u_\beta, \ldots, u_\kappa$ for any relabelling of the internal mesh points, giving a system of this type, with respect to which equation (43) is consistently ordered. To illustrate this point, we consider the iterative solution of equation (43) in some detail.

The boundary conditions given at the beginning of this section give $u_A = 15$, $u_B = 24$, $u_C = 21$, all other values on the boundary of the square being zero. The right-hand constant vector becomes $(15, 24, 21, 0, 0, \ldots, 0)'$, and the Jacobi process uses the recurrence relations

$$u_1^{(r+1)} = \tfrac{1}{4}(15 + u_2^{(r)} + u_4^{(r)})$$

$$u_2^{(r+1)} = \tfrac{1}{4}(24 + u_1^{(r)} + u_3^{(r)} + u_5^{(r)})$$

$$\cdots\cdots\cdots\cdots\cdots\cdots\cdots\cdots$$

$$u_9^{(r+1)} = \tfrac{1}{4}(0 + u_6^{(r)} + u_8^{(r)}).$$

There is no problem of ordering here, since the right-hand side of each relation involves values which all derive from the previous iteration. Any relabelling of the mesh points leads to a reordered system of recurrence relations such that the arithmetic used in calculating $\mathbf{u}^{(r+1)}$ from $\mathbf{u}^{(r)}$ is exactly the same.

With the Gauss-Seidel process (and the following discussion also applies in essence to S.O.R.) the situation is changed. The recurrence relations are now

$$u_1^{(r+1)} = \tfrac{1}{4}(15 + u_2^{(r)} + u_4^{(r)})$$

$$u_2^{(r+1)} = \tfrac{1}{4}(24 + u_1^{(r+1)} + u_3^{(r)} + u_5^{(r)})$$

$$\cdots\cdots\cdots\cdots\cdots\cdots\cdots\cdots$$

$$u_9^{(r+1)} = \tfrac{1}{4}(0 + u_6^{(r+1)} + u_8^{(r+1)}).$$

In terms of our fixed labelling, each iterative cycle involves the calculation of $(r + 1)$th approximations to u at the points $\alpha, \beta, \gamma, \delta, \epsilon, \zeta, \eta, \theta, \kappa$, in that order. If the internal mesh points are given a different numerical relabelling, the points are dealt with in a different order; for instance, the labelling of case (i) on p. 104 leads to the points being dealt with in the sequence $\alpha, \delta, \beta, \eta, \epsilon, \gamma, \theta, \zeta, \kappa$. In this new ordering, unlike the original, the value of $u_\delta^{(r+1)}$ is available and in use before $u_\beta^{(r+1)}$ is calculated. We may reasonably expect, therefore, that the two orderings will result in different values for $u_\beta^{(r+1)}$ and will have, in general, different arithmetical properties. However, it so happens that the equation giving $u_\beta^{(r+1)}$ in neither case involves u_δ, so that either ordering gives $u_\beta^{(r+1)}$ the same value. This proves to be the case for the entire set of $(r + 1)$th iterated values because equation (43) is consistently ordered with respect to the tridiagonal representation of case (i). The reader may like to confirm for himself that for both orderings the following table applies:

Component of $\mathbf{u}^{(r+1)}$ calculated	Component of $\mathbf{u}^{(r)}$ used	Component of $\mathbf{u}^{(r+1)}$ used
u_α	$u_\beta,\ u_\delta$	—
u_β	$u_\gamma,\ u_\epsilon$	u_α
u_γ	u_ζ	u_β
u_δ	$u_\epsilon,\ u_\eta$	u_α
u_ϵ	u_ζ, u_θ	$u_\beta,\ u_\delta$
u_ζ	u_κ	$u_\gamma,\ u_\epsilon$
u_η	u_θ	u_δ
u_θ	u_κ	$u_\epsilon,\ u_\eta$
u_κ	—	u_ζ, u_θ

On the other hand, the ordering of case (ii) involves the calculation, successively, of $(r + 1)$th approximations to u at the points $\alpha, \gamma, \epsilon, \eta, \kappa, \beta, \delta, \zeta, \theta$. Here the $(r + 1)$th iterated values of u at γ, ϵ, η and κ are all available before $u_\beta^{(r+1)}$ is calculated, and since the equation for $u_\beta^{(r+1)}$ actually involves u_γ and u_ϵ, the arithmetic cannot be the same in this instance. Then equation (43) is not consistently ordered with respect to the system of case (ii). However, although the two systems differ in the arithmetic details of their solution, their ultimate rates of convergence will be the same. This is so because their Jacobi matrices are similar (as shown on p. 89) and hence have the same set of eigenvalues; it follows that their S.O.R. matrices also have a common set of eigenvalues and, in consequence, the same spectral radii. The difference in arithmetic therefore does not affect the *ultimate* rate of convergence, though it will in general have some effect on the *average* rate of convergence.

For case (iii) the matrix has Property (A) but is inconsistently ordered. It may be shown (see, for instance, Varga [4] (p. 125)) that any inconsistent ordering of a block-tridiagonal system will have an inferior rate of convergence for Gauss-Seidel or S.O.R.

Finally, we consider convergence. Since the matrix of equation (43) has Property (A) and is consistently ordered, convergence of the Jacobi process implies convergence also of Gauss-Seidel and S.O.R. Gerschgorin's theorem shows that the eigenvalues of the matrix lie in the disc $|z - 4| \leqslant 4$ in the complex plane, centred at $z = 4$ and having radius 4. If we assume that the matrix is non-singular, so that $\lambda = 0$ is not an eigenvalue, and bear in mind that a real, symmetric matrix has only real eigenvalues, we conclude that the eigenvalues of our matrix all lie on the interval $0 < z \leqslant 4$ of the real axis. Then the matrix is positive-definite, and the Gauss-Seidel process will converge, as will also S.O.R.

If we wish to use S.O.R., we must now determine the optimum value for ω. For this particular problem (the solution of a five-point approximation to Laplace's equation over a square region, with values given on the boundary) it can be shown theoretically that the spectral radius of the Jacobi matrix is

$$\rho_J = \cos\frac{\pi}{p},*$$

where the original square region is subdivided into p^2 mesh squares. Our present example has $p = 4$, whence $\rho_J = \cos\frac{\pi}{4} = 0.7071$. It follows that $\rho_G = 0.5000$ and (from equations (27)) $\rho_S(\omega_0) = 0.1716$, where $\omega_0 = 1.1716$. Unfortunately, we can only calculate ρ_J and hence ω_0 in this simple manner for certain very restricted (albeit important) classes of problems. Usually, we must fall back on some numerical technique for estimating ω_0. The method suggested in Section 4.4.3 works well in the present example; when the Gauss-Seidel process is applied to equation (43) and the ratios of corresponding elements of successive displacement vectors $\mathbf{d}^{(r)} = \mathbf{u}^{(r+1)} - \mathbf{u}^{(r)}$ are calculated, we obtain the results

i	$\frac{d_i^{(2)}}{d_i^{(1)}}$	$\frac{d_i^{(3)}}{d_i^{(2)}}$	$\frac{d_i^{(4)}}{d_i^{(3)}}$	$\frac{d_i^{(5)}}{d_i^{(4)}}$	$\frac{d_i^{(6)}}{d_i^{(5)}}$	$\frac{d_i^{(7)}}{d_i^{(6)}}$	$\frac{d_i^{(8)}}{d_i^{(7)}}$	$\frac{d_i^{(9)}}{d_i^{(8)}}$	$\frac{d_i^{(10)}}{d_i^{(9)}}$
1	0.5250	0.4792	0.4511	0.4864	0.4965	0.4991	0.4998	0.4999	0.5000
2	0.3936	0.3503	0.4466	0.4851	0.4961	0.4990	0.4998	0.4999	0.5000
3	0.1779	0.3726	0.4573	0.4883	0.4970	0.4992	0.4998	0.5000	0.5000
4	1.1125	0.7149	0.5376	0.5087	0.5021	0.5005	0.5001	0.5000	0.5000
5	0.8333	0.5000	0.5000	0.5000	0.5000	0.5000	0.5000	0.5000	0.5000
6	0.3999	0.4687	0.4917	0.4979	0.4995	0.4999	0.5000	0.5000	0.5000
7	1.7000	0.8971	0.5553	0.5125	0.5030	0.5008	0.5002	0.5000	0.5000
8	1.2420	0.5747	0.5162	0.5039	0.5010	0.5002	0.5001	0.5000	0.5000
9	0.5662	0.5146	0.5035	0.5009	0.5002	0.5001	0.5000	0.5000	0.5000

The next iteration confirms these figures, and so 10 iterations have given ρ_G correct to four decimal places. At this stage, successive approximate solutions of the system are not yet agreeing to two decimal places. Our numerically determined value for ρ_G gives, as before, $\omega_0 = 1.1716$; with this value for the relaxation parameter, S.O.R. takes 10 iterations to produce two successive solution vectors which agree to four decimal places,

* If a rectangular region is subdivided into pq mesh squares, this result generalises to

$$\rho_J = \tfrac{1}{2}\left(\cos\frac{\pi}{p} + \cos\frac{\pi}{q}\right).$$

For further details, refer to Varga (4) (Chapter 6).

$$\mathbf{u} = \begin{pmatrix} 7.5000 \\ 11.4643 \\ 9.1071 \\ 3.5357 \\ 5.2500 \\ 3.9643 \\ 1.3929 \\ 2.0357 \\ 1.5000 \end{pmatrix}.$$

This is in agreement with the analytical solution of equation (43), which is

$$\mathbf{u} = \frac{1}{28}(210, 321, 255, 99, 147, 111, 39, 57, 42)'.$$

4.6 Further Developments

Much research has been devoted to further improvement of the methods of this chapter. We conclude with a brief sketch of two useful developments.

4.6.1 Non-stationary Processes

Earlier, we have shown that the rate of convergence of a process is related to $\rho(M)$, the spectral radius of its iteration matrix. On this basis, we have sought to minimise $\rho(M)$, in the case of S.O.R., by a suitable choice of the relaxation parameter ω. However, $\rho(M)$ does not give a measure of the overall rate of convergence of a process, but only of the *ultimate* or *asymptotic* rate of convergence, namely the rate at which the iterated solutions tend to the true solution once the process has settled down and its subsequent behaviour is determined entirely by the largest eigenvalue of its iteration matrix. It is usually possibly to obtain an improved *overall* or *average** rate of convergence by varying the relaxation parameter from one iteration to the next, the process now being non-stationary. In view of the complicated theory underlying the (stationary) S.O.R. process, the reader may anticipate, correctly, that the optimum choice of a *sequence* of varying relaxation parameters often poses very severe problems. Nevertheless, useful progress has been made in this direction.

The simplest and best-documented non-stationary process is that of Richardson. This is based on a formula derived from Jacobi's method in exactly the same way as S.O.R. is obtained from the Gauss-Seidel method:

$$\mathbf{x}^{(r+1)} = \{I - \alpha_r D^{-1}A\}\mathbf{x}^{(r)} + \alpha_r D^{-1}\mathbf{b}.$$

* Asymptotic and average rates of convergence are precisely defined by Varga [4] (pp. 62, 67), but we here seek to convey merely a rough idea of the distinction between them.

If the parameter α_r is treated as a constant, the formula may be optimised for best asymptotic rate of convergence, as was done earlier for S.O.R. Alternatively, if α_r varies with r we have a non-stationary process, for which the best average rate of convergence over k iterations is obtained on taking the α_r to be the reciprocals of the zeros of $T_k\left(\frac{b+a-2z}{b-a}\right)\Big/T_k\left(\frac{b+a}{b-a}\right)$. Here $T_k(z)$ is the Chebyshev polynomial of degree k and $a \leqslant z \leqslant b$ is the interval containing all the eigenvalues of $D^{-1}A$. Further details may be found in Forsythe & Wasow [1] (p. 226). A related but more general class of *semi-iterative* methods is discussed by Varga [4].

4.6.2 Block Relaxation

We will illustrate this concept by an example. Equation (43) may be written in the form

$$\begin{pmatrix} B & I & 0 \\ I & B & I \\ 0 & I & B \end{pmatrix}\begin{pmatrix} U_1 \\ U_2 \\ U_3 \end{pmatrix} = \begin{pmatrix} C_1 \\ C_2 \\ C_3 \end{pmatrix},$$

where

$$I = \begin{pmatrix} 1 & & \\ & 1 & \\ & & 1 \end{pmatrix}, \qquad B = \begin{pmatrix} -4 & 1 & 0 \\ 1 & -4 & 1 \\ 0 & 1 & -4 \end{pmatrix},$$

and the 9 x 1 column vectors have been partitioned as three 3 x 1 vectors. Equivalently,

$$\begin{aligned} BU_1 + U_2 \qquad &= C_1 \\ U_1 + BU_2 + U_3 &= C_2 \\ U_2 + BU_3 &= C_3 \end{aligned}$$

Here the unknowns are no longer single variables, as hitherto, but vectors composed of several variables. The *block Jacobi* method uses the vector iterative relations

$$\begin{aligned} BU_1^{(r+1)} &= C_1 - U_2^{(r)} \\ BU_2^{(r+1)} &= C_2 - U_1^{(r)} - U_3^{(r)} \\ BU_3^{(r+1)} &= C_3 - U_2^{(r)}, \end{aligned}$$

the utilisation of any one of which involves the solution of a tridiagonal system of scalar equations. In practice, the best method will usually be the direct technique given in Section 3.2.7 of Chapter 3. Failing any better information, the process could be started with

$$U_1^{(0)} = U_2^{(0)} = U_3^{(0)} = (0, 0, 0)'.$$

Analogously, we can define a *block Gauss-Seidel* and a *block S.O.R.* process, the latter being expressible as

$$U_1^{(r+1)} = U_1^{(r)} + \omega B^{-1}(C_1 - BU_1^{(r)} - U_2^{(r)})$$
$$U_2^{(r+1)} = U_2^{(r)} + \omega B^{-1}(C_2 - U_1^{(r+1)} - BU_2^{(r)} - U_3^{(r)})$$
$$U_3^{(r+1)} = U_3^{(r)} + \omega B^{-1}(C_3 - U_2^{(r+1)} - BU_3^{(r)}).$$

These methods may be applied in such a way that their requirements in terms of total arithmetic are precisely the same as those of the corresponding methods discussed earlier in this chapter, which should strictly be referred to as *point iterative* methods to distinguish them from their block iterative counterparts. Moreover, it can be shown that under suitable circumstances the block iterative methods have the faster convergence. Further details are given, with some comparative numerical results by Varga [4]. This reference also deals with *alternating direction implicit* (A.D.I.) methods, which are very powerful non-stationary block iterative techniques for solving linear systems associated with certain problems involving partial differential equations.

References

1. FORSYTHE, G. E. & WASOW, W. R., *Finite-Difference Methods for Partial Differential Equations*, Wiley (1960).
2. FOX, L., *Introduction to Numerical Linear Algebra*, O.U.P. (1964).
3. NATIONAL PHYSICAL LABORATORY, *Modern Computing Methods*, H.M. Stationery Office, (1961).
4. VARGA, R. S. *Matrix Iterative Analysis*, Prentice-Hall (1962).

Exercises

1. Perform several iterations of the Jacobi and Gauss-Seidel methods for the system

$$\begin{aligned} x_1 + 2x_2 - 2x_3 &= 1 \\ x_1 + x_2 + x_3 &= 0 \\ 2x_1 + 2x_2 + x_3 &= 1, \end{aligned}$$

and show that the former converges while the latter does not. Explain your findings by calculating the spectral radii of the relevant iteration matrices.

Prove that the Gauss-Seidel method converges when the first and third equations are interchanged.

2. Prove that, for the system

$$\begin{aligned} 2x_1 - x_2 + x_3 &= 1 \\ x_1 + x_2 + x_3 &= 6 \\ x_1 + x_2 - 2x_3 &= 0 \end{aligned}$$

the Gauss-Seidel method converges while the Jacobi does not.

3. Use (i) the Jacobi and (ii) the Gauss-Seidel method to solve

$$\begin{aligned} 10x_1 \qquad\qquad + x_3 &= 1 \\ 10x_2 + x_3 &= 2 \\ x_1 + x_2 + 10x_3 &= 1 \end{aligned}$$

to five significant figures. Find the value of ω giving the best rate of convergence for S.O.R.

4. Determine whether S.O.R. will converge for systems involving the following matrices:

$$\begin{pmatrix} 4 & -1 & -1 & -1 \\ -1 & 4 & -1 & -1 \\ -1 & -1 & 4 & -1 \\ -1 & -1 & -1 & 4 \end{pmatrix}, \quad \begin{pmatrix} 3 & 1 & 1 & 1 \\ 1 & 3 & 1 & 1 \\ 1 & 1 & 3 & 1 \\ 1 & 1 & 1 & 3 \end{pmatrix}, \quad \begin{pmatrix} 2 & 0 & -1 & -1 \\ 0 & 2 & -1 & -1 \\ -1 & 1 & 2 & 0 \\ -1 & -1 & 0 & 2 \end{pmatrix}.$$

5. Show that the matrix

$$\begin{pmatrix} 6 & -3 & -1 & 0 & -1 & 0 \\ -3 & 6 & 0 & -1 & 0 & -1 \\ -1 & 0 & 6 & -3 & 0 & -1 \\ 0 & -1 & -3 & 6 & -1 & 0 \\ -1 & 0 & 0 & -1 & 6 & -3 \\ 0 & -1 & -1 & 0 & -3 & 6 \end{pmatrix}$$

has Property (A), and obtain a block-tridiagonal reordering of it by the method of Section 4.4.3.

6. Prove that S.O.R. will converge for a system involving the reordered matrix of Exercise 5, and determine the best value to take for ω.

7. Perform 10 iterations in the Gauss-Seidel solution of the equations of Section 4.2.1, working to six decimal places. Use Aitken's δ^2-process to confirm the results quoted in Section 4.4.4 and to obtain similarly accelerated values for x_2, x_3 and x_4.

8. Show that the characteristic equations of the Jacobi and Gauss-Seidel iteration matrices for an $n \times n$ system $A\mathbf{x} = \mathbf{b}$ can be written as

$$\begin{vmatrix} \lambda a_{11} & a_{12} & \dots & a_{1n} \\ a_{21} & \lambda a_{22} & \dots & \lambda a_{2n} \\ \vdots & \vdots & & \vdots \\ a_{n1} & a_{n2} & \dots & \lambda a_{nn} \end{vmatrix} = 0, \qquad \begin{vmatrix} \lambda a_{11} & a_{12} & \dots & a_{1n} \\ \lambda a_{21} & \lambda a_{22} & \dots & a_{2n} \\ \vdots & \vdots & & \vdots \\ \lambda a_{n1} & \lambda a_{n2} & \dots & \lambda a_{nn} \end{vmatrix} = 0$$

respectively. Deduce that if $|A| \simeq 0$ (i.e. if A is nearly singular, or ill-conditioned), $\lambda \simeq 1$ is a root in either case, so that we can expect slow convergence at best.

9. By approximating the derivative, as shown in Section 4.5, express the differential equation $y'' - 25y = 50x$ as a linear algebraic equation in $y(x-h), y(x)$ and $y(x+h)$. Hence, given that $y(0) = 1$, $y(1) = -1$, find a system of equations whose solutions are approximations to y at $x = 0.2, 0.4, 0.6$ and 0.8.

10. Working to six decimal places, perform seven Gauss-Seidel iterations on the system of Exercise 9, and then estimate ρ_G by the method of Section 4.4.3. Hence determine ω_0, and complete the solution using S.O.R.

CHAPTER 5

ERRORS IN THE SOLUTION OF SETS OF EQUATIONS

5.1 Introduction

In this chapter, we shall consider the errors which may arise in solving a regular set of linear equations

$$A\mathbf{x} = \mathbf{b} \tag{1}$$

The problem of finding the inverse of a non-singular matrix A is in all respects similar to that of solving equations (1).

We first consider the effects of perturbations in the matrix A and in the vector $\mathbf{b}$. We shall examine the conditions under which such perturbations produce large errors in the result (the problem being ill-conditioned), and shall try to construct a measure of such ill-conditioning. We shall then show that the rounding errors introduced during the solution of the equations by Gaussian elimination, or by any similar method, is equivalent to a perturbation in the matrix A, and shall estimate bounds for the size of such a perturbation. Finally, we look at the effect of rounding errors on the iterative improvement of the solution of equations.

5.2 The Effect of Perturbations

5.2.1

Perturbations (i.e. small changes in input data) may seriously upset the solution of a set of equations. Suppose that we wish to solve the following set of equations:

$$\begin{aligned} 6x_1 + 4x_2 + 3x_3 &= b_1 \\ 20x_1 + 15x_2 + 12x_3 &= b_2 \\ 15x_1 + 12x_2 + 10x_3 &= b_3 \end{aligned} \tag{2}$$

where b_1, b_2, and b_3 are given as 13.1, 46.9, and 37.1 respectively. Without some prior analysis, it might be thought that rounding the values of the b's to two significant figures would not seriously affect the solution. However, the solution with the given right-hand side is

$$x_1 = 2.3, \quad x_2 = -3.7, \quad \text{and} \quad x_3 = 4.7$$

whereas the solution obtained if we round the right-hand side to

$$b_1 = 13, \quad b_2 = 47, \quad \text{and} \quad b_3 = 37,$$

becomes

$$x_1 = 1.0, \quad x_2 = 1.0, \quad \text{and} \quad x_3 = 1.0.$$

Clearly, the calculation is ill-conditioned, and the result obtained by rounding is entirely misleading. It is possible that the separate calculations that occur would arouse the suspicions of anyone of experience who was solving the equations by hand. For example, in the first line of a Gaussian elimination, we have the new coefficients of x_2 given by $15 - (20/6)4 = 5/3$, and $12 - (20/6)3 = 2$, involving differences between nearly equal numbers. But at no point is the cancellation very severe, and the extreme-ill-conditioning is not obvious.

5.2.2

We shall examine the matrix of coefficients in equation (2):

$$A = \begin{pmatrix} 6 & 4 & 3 \\ 20 & 15 & 12 \\ 15 & 12 & 10 \end{pmatrix} \tag{3}$$

the inverse of which is

$$A^{-1} = \begin{pmatrix} -6 & -4 & 3 \\ -20 & 15 & -12 \\ 15 & -12 & 10 \end{pmatrix}$$

The elements of A^{-1} are given by $M_{ij}/|A|$, where M_{ij} is a minor of A. Thus, the elements of A^{-1} tend to become indefinitely large as $|A|$ becomes small, and it is sometimes suggested that $|A|$ should be used as a measure of conditioning, the equations being taken to be ill-conditioned if the determinant is small. In this case, however, $|A| = 1$ and although this is perhaps smaller than one might expect from the magnitudes of the elements of A, it is not small enough to suggest serious ill-conditioning. Moreover, the elements of A^{-1} are not, in fact, large.

However, we can reasonably expect that the equations will be ill-conditioned if A is in some sense "nearly singular", even if we cannot use $|A|$ as a suitable indicator of this.

In the example, if we subtract from A the matrix

$$\Gamma_1 = \frac{15-\alpha}{112}\begin{pmatrix} 20 & -\alpha & 12 \\ -5\alpha & 56 & -3\alpha \\ 60 & -3\alpha & 36 \end{pmatrix} \approx \begin{pmatrix} 0.006 & -0.004 & 0.004 \\ -0.022 & 0.017 & -0.013 \\ 0.017 & -0.013 & 0.011 \end{pmatrix}, \tag{4}$$

where $\alpha = \sqrt{224}$, we obtain a singular matrix. The elements of Γ_1 are only about one thousandth of those of A and it is clearly reasonable to describe A as nearly singular. It is, of course, obvious from this that the solution of equations (2) is ill-conditioned with respect to perturbations of the coefficients as well as with respect to perturbations of the right-hand side.

If we now consider the matrix

$$A_\nu = A - \nu\Gamma_1 \tag{5}$$

we find its inverse to be

$$A_\nu^{-1} = A^{-1} - \frac{\nu}{1-\nu}\Gamma_2 \tag{6}$$

where

$$\Gamma_2 = \frac{15+\alpha}{15-\alpha}\Gamma_1 \approx \begin{pmatrix} 5.36 & -4.02 & 3.21 \\ -20.09 & 15 & -12.05 \\ 16.07 & -12.05 & 9.64 \end{pmatrix} \tag{7}$$

As ν approaches unity, the changes in A^{-1} due to the perturbations $\nu\Gamma_1$ in A become indefinitely large, but, even for modest values of ν, the term $\frac{\nu}{1-\nu}\Gamma_2$ changes the inverse beyond recognition.

If we take ν equal to $\frac{1}{2}$, so that the elements of A are perturbed by about half of 0.1 per cent, then the inverse of the perturbed matrix is

$$A_{\frac{1}{2}}^{-1} \approx \begin{pmatrix} 0.64 & 0.02 & -0.21 \\ 0.09 & 0 & 0.05 \\ -1.07 & 0.05 & 0.36 \end{pmatrix}$$

compared with the inverse of A itself,

$$A^{-1} = \begin{pmatrix} 6 & -4 & 3 \\ -20 & 15 & -12 \\ 15 & -12 & 10 \end{pmatrix}$$

The solution of equations (2), using the given right-hand values $b_1 = 13.1$, $b_2 = 46.9$, $b_3 = 37.1$, but with the perturbed matrix $A_{\frac{1}{2}}$, is approximately

$$x_1 = 1.53, \qquad x_2 = 3.03, \qquad x_3 = 11.01$$

instead of the required values

$$x_1 = 2.3, \qquad x_2 = -3.7, \qquad x_3 = 4.7.$$

5.2.3

In order to construct a suitable matrix Γ_1 for the purposes of the discussion in the preceding section, it was necessary to find out in which "direction" to change A so as to produce a large change in its inverse. This was done by examining the eigenvalues and eigenvectors of A. If the $(n \times n)$ matrix A has n linearly independent left-hand and right-hand eigenvectors $\mathbf{y}_i'$ and $\mathbf{x}_i$ and eigenvalues λ_i, then A can be written in the form

$$A = \sum_{i=1}^{n} \lambda_i \mathbf{x}_i \mathbf{y}_i' \tag{8}$$

Also, the inverse of A is given by

$$A^{-1} = \sum_{i=1}^{n} \left(\frac{1}{\lambda_i}\right) \mathbf{x}_i \mathbf{y}_i' \tag{9}$$

If the eigenvalues are arranged in order of absolute magnitude, so that

$$|\lambda_1| \leqslant |\lambda_2| \leqslant \ldots \leqslant |\lambda_n|$$

and if $|\lambda_1|$ is small compared with $|\lambda_n|$, then the term $\lambda_1 \mathbf{x}_1 \mathbf{y}_1'$ in A is small compared with the term $\lambda_n \mathbf{x}_n \mathbf{y}_n'$, and contributes little towards the elements of A. On the other hand, the corresponding term $(1/\lambda)\mathbf{x}_1 \mathbf{y}_1'$ is an important constituent of the inverse A^{-1}. In the example considered, A has eigenvalues and corresponding eigenvectors as follows:

$$\lambda_1 = 15 - \alpha; \qquad \mathbf{x}_1 = \begin{pmatrix} 4/\alpha\sqrt{2} \\ -1/\sqrt{2} \\ 12/\alpha\sqrt{2} \end{pmatrix}, \qquad \mathbf{y}_1' = \left(\frac{20}{\alpha\sqrt{2}}, \ -\frac{1}{\sqrt{2}}, \ \frac{12}{\alpha\sqrt{2}}\right),$$

$$\lambda_2 = 1; \qquad \mathbf{x}_2 = \begin{pmatrix} 3/\sqrt{14} \\ 0 \\ -5/\sqrt{14} \end{pmatrix}, \qquad \mathbf{y}_2' = \left(\frac{3}{\sqrt{14}}, \ 0, \ -\frac{1}{\sqrt{14}}\right),$$

$$\lambda_3 = 15 + \alpha; \qquad \mathbf{x}_3 = \begin{pmatrix} 4/\alpha\sqrt{2} \\ 1/\sqrt{2} \\ 12/\alpha\sqrt{2} \end{pmatrix}, \qquad \mathbf{y}_3' = \left(\frac{20}{\alpha\sqrt{2}}, \ \frac{1}{\sqrt{2}}, \ \frac{12}{\alpha\sqrt{2}}\right),$$

where, as before, $\alpha = \sqrt{224}$. The matrix Γ_1 was chosen as $\lambda_1 \mathbf{x}_1 \mathbf{y}_1'$. The extreme ill-conditioning goes together with the large ratio of largest to smallest eigenvalue, viz:— $\dfrac{15 + \alpha}{15 - \alpha} \approx 900.$

We can use the same analysis of the matrix of coefficients to examine the susceptibility of the solution to perturbations in the right-hand vector. To exhibit the ill-conditioning, we should choose **b** so that the solution **x** is comparatively small, and then choose the perturbation, $\delta\mathbf{b}$, so that the resulting error, $\delta\mathbf{x}$, is large. We can achieve this by taking **b** parallel to the column eigenvector corresponding to the largest eigenvalue, say $\mathbf{b} = b\,\mathbf{x}_n$, and $\delta\mathbf{b}$ parallel to the column eigenvector corresponding to the smallest eigenvalue, say $\delta\mathbf{b} = \delta\mathbf{x}_1$. We have,

$$\mathbf{x} = A^{-1}\mathbf{b} = b\sum_{i=1}^{n}\left(\frac{1}{\lambda_i}\right)\mathbf{x}_i\mathbf{y}_i'\mathbf{x}_n = \left(\frac{b}{\lambda_n}\right)x_n = \left(\frac{1}{\lambda_n}\mathbf{b},\right)$$

since $\mathbf{y}_i'\mathbf{x}_k = 0$ for $i \neq n$, and $y_n'\mathbf{x}_n = 1$.

Also,

$$\delta\mathbf{x} = A^{-1}\delta\mathbf{b} = \delta\sum_{i=1}^{n}\left(\frac{1}{\lambda_i}\right)\mathbf{x}_i\mathbf{y}_i'\mathbf{x}_1 = \left(\frac{\delta}{\lambda_1}\right)\mathbf{x}_1 = \left(\frac{1}{\lambda_1}\right)\delta\mathbf{b}.$$

The ratio of the norm of $\delta\mathbf{x}$ to that of **x** itself is

$$\frac{\|\delta\mathbf{x}\|}{\|\mathbf{x}\|} = \frac{\|\delta\mathbf{b}\|}{|\lambda_1|}\cdot\frac{|\lambda_n|}{\|\mathbf{b}\|} = \frac{|\lambda_n|}{|\lambda_1|}\cdot\frac{\|\delta\mathbf{b}\|}{\|\mathbf{b}\|}.$$

The relative perturbation is magnified by the factor $|\lambda_n|/|\lambda_1|$, which may be large, and was in fact about 900 in the example considered. If, in this example, we choose $\mathbf{b} = \begin{pmatrix}4\\15\\12\end{pmatrix}$, (which is very nearly parallel to the eigenvector $\mathbf{x}_3$ corresponding to the largest eigenvalue), then $\mathbf{x} = \begin{pmatrix}0\\-1\\0\end{pmatrix}$. If also $\delta\mathbf{b}$ is chosen to be $\begin{pmatrix}0.004\\-0.015\\0.012\end{pmatrix}$, (which is nearly parallel to the eigenvector corresponding to the smallest eigenvalue), then $\delta\mathbf{x} = \begin{pmatrix}0.12\\-0.449\\0.360\end{pmatrix}$. The magnification of error is given by

$$\frac{\|\delta\mathbf{x}\|}{\|\mathbf{x}\|}\bigg/\frac{\|\delta\mathbf{b}\|}{\|\mathbf{b}\|} \approx \left(\frac{0.59}{1}\right)\bigg/\left(\frac{\sqrt{405}\times 10^{-3}}{\sqrt{405}}\right) = 590.$$

5.3 Condition Number

5.3.1

In the preceding section we saw that the solution of

$$A\mathbf{x} = \mathbf{b}$$

was highly sensitive to perturbations if the ratio of largest to smallest eigenvalue was large. It was not, however, shown (or suggested) that the magnification of error is bounded by this ratio of eigenvalues. In this section, we construct such a bound.

5.3.2

Suppose that the matrix of coefficients, A, and the vector $\mathbf{b}$ in equation (1), are subject to perturbations δA and $\delta \mathbf{b}$ respectively, and that as a consequence the true solution, $\mathbf{x}$, is in error by the vector $\delta \mathbf{x}$. We have

$$(A + \delta A)(\mathbf{x} + \delta \mathbf{x}) = \mathbf{b} + \delta \mathbf{b}.$$

Subtracting equation (1) from this we obtain

$$A\delta\mathbf{x} + \delta A\mathbf{x} + \delta A\delta\mathbf{x} = \delta\mathbf{b},$$

so that,

$$\delta\mathbf{x} = A^{-1}\{\delta\mathbf{b} - \delta A\mathbf{x} - \delta A\delta\mathbf{x}\}.$$

It follows that, for any choice of norm,

$$\|\delta\mathbf{x}\| \leqslant \|A^{-1}\| \{\|\delta\mathbf{b}\| + \|\delta A\| \|\mathbf{x}\| + \|\delta A\| \|\delta\mathbf{x}\|\} \tag{10}$$

Also, from equation (1),

$$\|\mathbf{b}\| \leqslant \|A\| \|\mathbf{x}\|$$

so that

$$\frac{1}{\|\mathbf{x}\|} \leqslant \frac{\|A\|}{\|\mathbf{b}\|}.$$

Then, from equation (10),

$$(1 - \|A^{-1}\| \|\delta A\|)\frac{\|\delta\mathbf{x}\|}{\|\mathbf{x}\|} \leqslant \|A^{-1}\| \left\{\frac{\|\delta\mathbf{b}\|}{\|\mathbf{x}\|} + \|\delta A\|\right\} \leqslant \|A^{-1}\| \left\{\frac{\|A\| \|\delta\mathbf{b}\|}{\|\mathbf{b}\|} + \|\delta A\|\right\} \tag{11}$$

The inequality (11) only affords a useful bound for $\|\delta\mathbf{x}\|/\|\mathbf{x}\|$ when $\|A^{-1}\| \|\delta A\|$ is less than unity, and in this case we get

$$\frac{\|\delta\mathbf{x}\|}{\|\mathbf{x}\|} \leqslant \frac{K}{1 - K \|\delta A\|/\|A\|} \left\{\frac{\|\delta\mathbf{b}\|}{\|\mathbf{b}\|} + \frac{\|\delta A\|}{\|A\|}\right\} \tag{12}$$

where $K = \|A\| \|A^{-1}\|$. (13)

Clearly K may be used as a condition number. To have confidence in a solution we require $K\|\delta A\|/\|A\|$ to be at least fairly small. We can then replace the divisor, $(1 - K\|\delta A\|/\|A\|)$, by unity, and so K provides a bound to the magnification of the perturbation in $\mathbf{b}$, and (approximately) to that of the perturbation in A also.

5.3.3

A similar analysis applies to the problem of finding the inverse of a matrix. If X denotes the inverse of A, and δK the error in X due to a perturbation δA in A, then

$$\frac{\|\delta X\|}{\|X\|} \leqslant \frac{K\|\delta A\|/\|A\|}{1-K\|\delta A\|/\|A\|}$$

so that K affords a measure of conditioning for both problems.

Note that, for any norm except the Euclidean, the norm of a unit matrix is unity so that

$$K = \|A\|\,\|A^{-1}\| \geqslant \|AA^{-1}\| \geqslant \|I\| = 1.$$

5.3.4

The following table shows the value of the condition number, K, for each of the four norms, for the matrix of the previous section:

Type of norm	$\|A\|$	$\|A^{-1}\|$	K
ℓ_1	41	41	1681
ℓ_∞	47	47	2209
Euclidean	36	36	1299
ℓ_2	36	36	1296

(The values in the last two lines are approximate, and the Euclidean and ℓ_2 norms are not, in fact, exactly equal.) We showed that it was possible to obtain a magnification of the perturbations in b of about 900. We shall now show that it can actually be equal to K.

Given a matrix M then, for any norm other than the Euclidean, it is always possible to choose a vector $\mathbf{v}$ such that

$$\|M\mathbf{v}\| = \|M\|\,\|\mathbf{v}\|.$$

(see Appendix). If we choose $\mathbf{b}$, and hence $\mathbf{x}$, so that

$$\|\mathbf{b}\| = \|A\mathbf{x}\| = \|A\|\,\|\mathbf{x}\|,$$

and choose $\delta\mathbf{b}$ so that

$$\|\delta\mathbf{x}\| = \|A^{-1}\delta\mathbf{b}\| = \|A^{-1}\|\,\|\delta\mathbf{b}\|,$$

then we have

$$\frac{\|\delta\mathbf{x}\|}{\|\mathbf{x}\|} = \|A^{-1}\|\,\|\delta\mathbf{b}\|\frac{\|A\|}{\|\mathbf{b}\|} = K\frac{\|\delta\mathbf{b}\|}{\|\mathbf{b}\|}.$$

Illustrating this with the same matrix as before:

$$A = \begin{pmatrix} 6 & 4 & 3 \\ 20 & 15 & 12 \\ 15 & 12 & 10 \end{pmatrix}, \qquad A^{-1} = \begin{pmatrix} 6 & -4 & 3 \\ -20 & 15 & -12 \\ 25 & -12 & 10 \end{pmatrix}$$

and using the ℓ_∞ norm for maximum effect, we take

$$\mathbf{x} = \begin{pmatrix} 1 \\ 1 \\ 1 \end{pmatrix} \quad \text{so that} \quad \mathbf{b} = \begin{pmatrix} 13 \\ 47 \\ 37 \end{pmatrix},$$

and, for some small quantity ν, $\delta\mathbf{b} = \nu \begin{pmatrix} 1 \\ -1 \\ 1 \end{pmatrix}$ ao that $\delta\mathbf{x} = \nu \begin{pmatrix} 13 \\ 47 \\ 37 \end{pmatrix}$

Then: $$\frac{\|\delta\mathbf{x}\|}{\|\mathbf{x}\|} \bigg/ \frac{\|\delta\mathbf{b}\|}{\|\mathbf{b}\|} = \frac{47\,|\nu|}{1} \bigg/ \frac{|\nu|}{47} = 2209 = K.$$

Note that finding the vector $\mathbf{b}$ which will give the worst case is not difficult for the ℓ_1 and ℓ_∞ norms:

(i) Let the row of A which has the largest sum of moduli be row α. Choose the elements of $\mathbf{x}$ by writing

$$x_j = a_{\alpha j}/|a_{\alpha j}| = \text{sgn.}(a_{\alpha j}) = \pm 1$$

if $a_{\alpha j} \neq 0$, and $x_j = 1$ if $a_{\alpha j} = 0$. Then $\mathbf{b}$ is given by

$$b_i = \sum_{j=1}^{n} a_{ij} x_j = \sum_{j=1}^{n} a_{ij}\, a_j/|a_{\alpha j}|.$$

Hence, $\|\mathbf{b}\|_\infty = |b_\alpha| = \sum_{j=1}^{n} |a_{\alpha j}| = \|A\|_\infty$, and $\|\mathbf{x}\|_\infty = 1$,

and so,

$$\|\mathbf{b}\|_\infty = \|A\|_\infty \|\mathbf{x}\|_\infty .$$

(ii) Let the column of A which has the largest sum of moduli be column β. Choose $x_j = 0$ for $j \neq \beta$, and $x_\beta = 1$.

Then, $$b_i = \sum_{j=1}^{n} a_{ij} x_j = a_{i\beta}$$

and so

$$\|\mathbf{b}\|_1 = \sum_{i=1}^{n} |b_i| = \sum_{i=1}^{n} |a_{i\beta}| = \|A\|_1 .$$

Since $\|\mathbf{x}\|_1 = 1$, we again have $\|\mathbf{b}\|_1 = \|A\|_1 \|\mathbf{x}\|_1$.

Using ℓ_1 instead of ℓ_∞ in the above example would select **b** as being either $\begin{pmatrix} 6 \\ 20 \\ 15 \end{pmatrix}$ or $\begin{pmatrix} 4 \\ 15 \\ 12 \end{pmatrix}$ instead of $\begin{pmatrix} 13 \\ 47 \\ 37 \end{pmatrix}$, and these three vectors are nearly parallel.

5.3.5

If **b** is perpendicular to the worst case vector, the errors will be much less severe, even if $\delta\mathbf{b}$ turns out to be the worst possible. In the example, take **b** as the vector $\begin{pmatrix} 9 \\ -8 \\ 7 \end{pmatrix}$. This is perpendicular both to $\begin{pmatrix} 4 \\ 15 \\ 12 \end{pmatrix}$ and to $\begin{pmatrix} 13 \\ 47 \\ 37 \end{pmatrix}$ and

$\mathbf{x} = \begin{pmatrix} 107 \\ -384 \\ 301 \end{pmatrix}$. Now $\|\mathbf{b}\|_\infty = 9$, and $\|\mathbf{x}\|_\infty = 384$, and with the worst case,

$\delta\mathbf{b} = \nu \begin{pmatrix} 1 \\ -1 \\ 1 \end{pmatrix}$, we have: $\dfrac{\|\delta\mathbf{x}\|_\infty}{\|\mathbf{x}\|_\infty} \Big| \dfrac{\|\delta\mathbf{b}\|_\infty}{\|\mathbf{b}\|_\infty} = \dfrac{47\nu}{384} \Big| \dfrac{\nu}{9} \approx 1.1.$

This compares with 2,209 for the worst possible **b**. Similar results hold for other norms.

5.3.6

If A is a symmetric matrix, then its ℓ_2 norm is equal to the eigenvalue of largest modulus of A. Let the eigenvalues of A be $\lambda_1, \lambda_2, \ldots, \lambda_n$, where $|\lambda_1| \leqslant |\lambda_2| \leqslant \ldots \leqslant |\lambda_n|$. Then A^{-1} is symmetric, and its eigenvalues are the reciprocals of those of A. Hence,

$$\|A\|_2 = |\lambda_n| \qquad \text{and} \qquad \|A^{-1}\|_2 = 1/|\lambda_1|$$

so that,

$$K = |\lambda_n|/|\lambda_1|.$$

For symmetric matrices, the condition number based on the ℓ_2 norm is the same as the measure of ill-conditioning obtained in the previous section.

5.4 Rounding Errors in the Solution of Equations

5.4.1

As pointed out in Chapter 2, it is difficult, if not impossible, to construct useful bounds for the effects of rounding errors in an extended piece of calculation by tracing through that calculation. Instead, we use backward analysis and equate the effected of rounding errors to the effects of an equivalent initial perturbation.

That rounding errors can be serious may be seen by examining the effect of rounding the coefficients of the equations, the right-hand side values and the subsequent arithmetic being supposed exact.

5.4.2 The Effect of Rounding Coefficients

If the coefficients are stored in floating-point form with a t-bit mantissa, then the rounding errors δa_{ij} are bounded by

$$|\delta a_{ij}| \leqslant 2^{-t} |a_{ij}|.$$

Using the ℓ_1, ℓ_∞, or Euclidean norms, this gives $\|\delta A\| \leqslant 2^{-t}\|A\|$, but with the ℓ_2 norm the best we can say is that

$$\|\delta A\|_2 \leqslant \sqrt{n} \,.\, 2^{-t}\|A\|_2.$$

Using the ℓ_2 norm, and the results of the previous section,

$$\frac{\|\delta \mathbf{x}\|_2}{\|\mathbf{x}\|_2} < \frac{2^{-t}\sqrt{n}K}{1-2^{-t}\sqrt{n}K},$$

and to be sure that the effect of rounding errors is not serious we require that $2^{-t}\sqrt{n}K$ should be small compared with unity. Hence,

$$K = \|A\|_2 \|A^{-1}\|_2 \ll 2^t/\sqrt{n}.$$

This restriction is not negligible if the equations are ill-conditioned. For example, if A is the (5×5) Hilbert matrix

$$a_{ij} = \frac{10}{i+j}, \quad (i, j = 1, 2, \ldots, 5),$$

then $\|A\|_2 = 10.56$, and $\|A^{-1}\|_2 = \frac{1}{7} \times 10^6$, so that

$$K \approx 1.5 \times 10^6.$$

To be sure that the effect of rounding the coefficients is negligible we require

$$2^t \gg 1.5 \times 10^6\sqrt{5} \approx 3.3 \times 10^6 \approx 2^{21}.$$

We conclude that if the computer works to only about 20-bit accuracy the solution may be hopelessly inaccurate, even without considering any errors after the initial rounding. The same conclusion follows if the computer works exactly but the coefficients are entered into the computer as six decimal digit numbers instead of as ratios of integers.

5.4.3 Rounding Errors in Gaussian Elimination

Consider now the reduction of the equations to triangular form by Gaussian elimination. If we write the equations to be solved as

$$A^{(0)}\mathbf{x} = \mathbf{b}^{(0)} \tag{14}$$

then the elimination consists of transforming these equations successively into the sets of equations:

$$A^{(k)}\mathbf{x} = \mathbf{b}^{(k)}, \qquad k = 1, 2, \ldots, n-1,$$

in such a way that $A^{(k)}$ has zeros below the diagonal in the first k columns; hence, finally, $A^{(n-1)}$ is upper triangular. $A^{(k+1)}$ is obtained from $A^{(k)}$ by subtracting from row i, $(i = k+2, \ldots, n)$, the expression $m_{i,k+1} \times$ row $(k+1)$, where $m_{i,k+1} = 1\, a^{(k)}_{i,k+1}/a^{(k)}_{k+1,k+1}$.

Consider first the effect of rounding errors on the elements of $A^{(n-1)}$. The only calculated (non-zero) elements of $A^{(n-1)}$ are in the upper triangle, and fixing on a particular element in this upper triangle we consider the successive values of the elements in that position in matrices $A^{(0)}, A^{(1)}$, and so on. That is, we consider the successive values $a^{(1)}_{ij}, a^{(2)}_{ij}, \ldots, a^{(n-1)}_{ij}$, when $j \geqslant i$. However, it is easily seen that the element $a^{(k)}_{ij}$ remains constant for $k \geqslant i-1$, and therefore we need only look at the sequence of values as far as $a^{(i-1)}_{ij}$.

In what follows all a's and m's refer to computed values, and so we shall not need a special notation to distinguish ideal from computed values. Assuming the use of floating-point arithmetic with a t-bit mantissa,

$$a^{(k+1)}_{ij} = a^{(k)}_{ij}\,(1 + \alpha^{(k)}_{ij}) - m_{i,k+1}a^{(k)}_{k+1,j}\,(1 - \beta^{(k)}_{ij}) \tag{15}$$

where

$$|\alpha^{(k)}_{ij}| \leqslant 2^{-t} \qquad \text{and} \qquad |\beta^{(k)}_{ij}| \leqslant 2 \times 2^{-t}, \qquad \text{for } k = 0, 1, \ldots, i-2.$$

Adding together these $(i-1)$ equations yields

$$\sum_{k=0}^{i-2} a^{(k+1)}_{ij} = \sum_{k=0}^{i-2} a^{(k)}_{ij} - \sum_{k=0}^{i-2} m_{i,k+1}a^{(k)}_{k+1,j} + \epsilon_{ij},$$

which, upon cancellation of terms in the first two sums, becomes

$$a^{(i-1)}_{ij} + \sum_{k=0}^{i-2} m_{i,k+1}a^{(k)}_{k+1,j} = a^{(0)}_{ij} + \epsilon_{ij}, \qquad (\text{for } j \geqslant i) \tag{16}$$

where

$$\epsilon_{ij} = \sum_{k=0}^{i-2}\left[\alpha^{(i)}_{ij}\,\alpha^{(k)}_{ij} - \beta^{(k)}_{ij}\, m_{i,k+1}a^{(k)}_{k+1,j}\right] \tag{17}$$

Equations (16) define the reduction of a set of equations

$$(A + E)\mathbf{x} = \mathbf{b}$$

with coefficients $(a^{(0)}_{ij} + \epsilon_{ij})$. That is, the rounding errors which occur in the calculation of the lower triangular matrix are equivalent to a perturbation of the original matrix with elements in the upper triangle given by equation (17). (Note that the $m_{i,k+1}$ and the $a_{ij}^{(k)}$ which occur in equation (17) are not exactly those which would occur in the reduction of these perturbed equations; however, assuming that they differ from them by terms of order 2^{-t}, we may neglect these second order errors.)

5.4.4

It remains to examine the calculation of the multipliers m_{ij}, which only exist for $i > j$. We shall see that the errors arising are equivalent to a perturbation in the lower triangle of the original matrix of coefficients. Considering the successive values of $a_{ij}^{(k)}$, where $i > j$, we have initially the same eliminations as before:

$$a_{ij}^{(k+1)} = a_{ij}^{(k)} (1 + \alpha_{ij}^{(k)}) - m_{i,k+1} a_{k+1,j}^{(k)} (1 - \beta_{ij}^{(k)}) \tag{18}$$

for $k = 0, 1, \ldots, j-2$.

In the next step, the element obtained is used to compute $m_{i,j}$. We have,

$$m_{ij} = \left[a_{ij}^{(j-1)} / a_{jj}^{(j-1)}\right] (1 + \gamma_{ij}), \qquad \text{where } |\gamma_{ij}| \leqslant 2^{-t}.$$

Re-writing this in the form

$$0 = a_{ij}^{(j-1)} - m_{i,j} a_{jj}^{(j-1)} + \gamma_{ij} a_{ij}^{(j-1)}$$

and adding this equation together with the $(k-2)$ equations in (18), we obtain, after cancellation and re-arrangement

$$\sum_{k=0}^{j-1} m_{i,k+1} a_{k+1,j}^{(k)} = a_{ij}^{(0)} + \epsilon_{ij}, \qquad (\text{for } i > j), \tag{19}$$

where
$$\epsilon_{ij} = \sum_{k=0}^{j-2} \left[\alpha_{ij}^{(i)} a_{ij}^{(k)} + \beta_{ij}^{(k)} m_{i,k+1} a_{k+1,j}^{(k)}\right] + \gamma_{ij} a_{ij}^{(j-1)}. \tag{20}$$

Thus, the rounding errors in the calculation of the multipliers, m_{ij}, are equivalent to a perturbation in the lower triangle given by (20). Bounds on the elements of the perturbation matrix $E = (e_{ij})$ are given by

$$\left.\begin{aligned} |\epsilon_{ij}| &\leqslant 2^{-t} \sum_{k=0}^{i-2} \{|a_{ij}^{(k)}| + 2|m_{i,k+1}|\,|a_{k+1,j}^{(k)}|\}, && \text{for } i \leqslant j, \\ |\epsilon_{ij}| &\leqslant 2^{-t}\left[\sum_{k=0}^{j-2} \{|a_{ij}^{(k)}| + 2|m_{i,k+1}|\,|a_{k+1,j}^{(k)}|\} + |a_{ij}^{(j-1)}|\right], && \text{for } i > j \end{aligned}\right\} \tag{21}$$

Note that if we assume that all the rounding errors may be the worst possible, we cannot write smaller bounds than this. To keep the equivalent perturbations relatively small, we require all the m_{ij} and $a_{ij}^{(k)}$ not to be large (compared with the original coefficients $a_{ij}^{(o)}$). Even with exact arithmetic it is, of course necessary to be prepared to re-order the original equations to avoid trying to use as pivot an element with value zero (corresponding to infinitely large multipliers $m_{i,k}$). Partial pivoting (*q.v.*) not only avoids a breakdown in the algorithm but further ensures that all the $m_{i,k}$ will be less than one in modulus. At the same time, it ensures that the elements $a_{ij}^{(k)}$ do not grow without bound. If M denotes the largest element, in modulus, in the original matrix of coefficients, then, with partial pivoting, we get

$$|a_{ij}^{(k)}| \leqslant 2^k M \text{ for } k \leqslant i-1, \quad \text{and} \quad |a_{ij}^{(k)}| \leqslant 2^{i-1} M \text{ for } k \geqslant i. \tag{22}$$

Using these bounds in (21), with $|m_{i,k}| \leqslant 1$, we obtain

$$\left.\begin{aligned} &|\epsilon_{ij}| \leqslant 3(2^{i-1}-1)2^{-t}M \text{ for } i \leqslant j, \\ \text{and } &|\epsilon_{ij}| \leqslant (2^{j+1}-3)2^{-t}M \text{ for } i > j \end{aligned}\right\} \tag{23}$$

For the ℓ_1 norm of E we have,

$$\|E\|_1 = \max_{1 \leqslant j \leqslant n} \sum_{i=1}^{n} |\epsilon_{ij}|$$

$$\leqslant \max_{1 \leqslant j \leqslant n} \left\{ \sum_{i=1}^{j} 3(2^{i-1}-1)2^{-t}M + \sum_{i=j+1}^{n} (2^{j+1}-3)2^{-t}M \right\}$$

The maximum occurs when $j = n$ and we find

$$\|E\|_1 = 3(2^n - n - 1)2^{-t}M. \tag{24}$$

$$\text{Similarly, } \|E\|_\infty = \max_{1 \leqslant i \leqslant n} \sum_{j=1}^{n} |\epsilon_{ij}|$$

$$\leqslant \max_{1 \leqslant i \leqslant n} \left\{ \sum_{j=1}^{i-1} (2^{j+1}-3)2^{-t}M + \sum_{j=i}^{n} 3(2^{i-1}-1)2^{-t}M \right\}$$

$$= (7.2^{n-1} - 3n - 4)2^{-t}M. \tag{25}$$

Using these bounds, we should require that the number of equations should certainly be less than the numbers of digits used to store the mantissa in the floating-point binary arithmetic used, even if the equations are well-conditioned.

For both ℓ_1 and ℓ_∞ norms, the norm of the matrix of coefficients satisfies the inequalities

$$M \leqslant \|A\| \leqslant nM$$

so that, without increasing the bounds given in (24) and (25) by much, we can write for both norms

$$\frac{\|E\|}{\|A\|} \leqslant 7.2^{n-1}.2^{-t}.$$

The resultant error in the solution of the equations is then bounded by

$$\frac{\|\delta \mathbf{x}\|}{\|\mathbf{x}\|} \leqslant K.7.2^{n-1}.2^{-t}. \tag{26}$$

Since K is never less than unity, we should require (even for a perfectly conditioned set of equations) that 7.2^{n-1-t} should be small. With a typical value of about 30 for t we should thus be led to doubt the possibility of solving a set of more than about 20 equations.

5.4.5

In the early days of automatic digital computers, it was, in fact, held that without double length arithmetic rounding errors would prohibit the use of Gaussian elimination methods for the solution of large sets of equations. Largely as the result of the work of Dr. Wilkinson, it is now realised that the errors arising in such methods are likely to be very much smaller than had been feared.

Following Wilkinson (*Rounding Errors in Algebraic Processes*, H.M.S.O., 1963) we let g denote the maximum value, for all i, j, and k, of $|a_{ij}^{(k)}|$. Then, with $|m_{i,j}| \leqslant 1$, we obtain from (21),

$$\begin{aligned} &|\epsilon_{ij}| \leqslant 3(i-1)g\,2^{-t}, \quad &&\text{for } i \leqslant j \\ \text{and} \quad &|\epsilon_{ij}| \leqslant (3j-2)g\,2^{-t}, \quad &&\text{for } i > j \end{aligned} \tag{27}$$

Then,

$$\begin{aligned} &\|E\|_1 \leqslant 3n(n-1)g\,2^{-t} \\ \text{and} \quad &\|E\|_\infty \leqslant (3n-n-2)g\,2^{-t} \end{aligned} \tag{28}$$

With partial pivoting, we can only guarantee that $g \leqslant 2^{n-1}M$. Substituting this in (28) yields worse bounds than were obtained previously. Nevertheless, Wilkinson states as a fact of experience it is extremely rare for g to exceed $4M$. Moreover, if the equations are ill-conditioned, the $a_{ij}^{(k)}$ generally decrease as k increases so that g will often be equal to M. We may expect, then, that $\|E\|/\|A\|$ will be only a small multiple of $n^2 2^{-t}$. Taking $t = 30$, we may expect rounding errors in the triangulation not to be serious provided n^2 is small compared with 2^{30}, i.e. provided n is not greater than about 200.

Note that the growth of elements at the maximum rate shown by (22) is possible with partial pivoting, so that the bounds given in (24) and (25) are attainable (even if unlikely to occur). It is possible to guarantee a slower rate of growth of the $a_{ij}^{(k)}$ by using complete pivoting (Wilkinson: "Error Analysis of Direct Methods of Matrix Inversion", *J. Assoc. Comp. Math* 8, 1961, pp. 281–330) but in the light of the previous remarks about the maximum value of g it is usually not felt to be worth the extra computing time necessary. This possible decrease in the rate of growth is the only advantage of complete pivoting over partial pivoting. It is, of course, easy to keep check on the magnitudes of the $a_{ij}^{(k)}$

as they are computed, and to introduce complete pivoting if they appear to be growing too rapidly.

5.4.6

The first part of the Gaussian process is equivalent to factorising the matrix of coefficients A (or $A + E$ in the presence of rounding errors) into the product of a lower triangular matrix L and an upper triangular matrix U:

$$L U \mathbf{x} = \mathbf{b} \tag{29}$$
$$\text{where } LU = A + E.$$

The elements of L are equal to the multipliers m_{ij} below the main diagonal and equal to unity on the diagonal; the elements of U are the final values of the $a_{ij}^{(k)}$ in the upper triangle. Thus,

$$\left.\begin{aligned} &|\ell_{ij}| \leqslant 1 && \text{for } j \leqslant i,\\ \text{and}\quad &|u_{ij}| \leqslant 2^{i-1} M && \text{for } j \geqslant i, \text{ (from (22))} \end{aligned}\right. \tag{30}$$

We examine the effect of rounding errors which occur while solving in turn the sets of equations

$$\left.\begin{aligned} L\mathbf{y} &= \mathbf{b}\\ U\mathbf{x} &= \mathbf{y} \end{aligned}\right\} \tag{31}$$

Without rounding errors, the elements of $\mathbf{y}$ are

$$y_i = \left\{ b_i - \sum_{j=1}^{i-1} \ell_{ij} y_j \right\} \bigg/ \ell_{ii}.$$

Using floating point arithmetic with a t-bit mantissa, we may write

$$y_i = \left\{ (1 + \phi_{ii}) b_i - \sum_{j=1}^{i-1} \ell_{ij} y_j (1 + \theta_{ij})(1 + \phi_{ij}) \right\} (1 + \nu_i)/\ell_{ii} \tag{32}$$

where $\quad |\theta_{ij}| \leqslant 2^{-t}, \quad |\nu_i| \leqslant 2^{-t}, \quad$ and $\quad |\phi_{ij}| \leqslant (i - j + 1) 2^{-t}$.

(The θ's are due to rounded multiplication, the ν's to rounded division, and the ϕ's to rounded addition. It is assumed that the sum has been evaluated first and then subtracted from b_i.) In Fact, with Gaussian elimination, the ℓ_{ii} are all exactly unity and so may be omitted, together with the associated errors ν_i. We let them stand in (32) so that the analysis remains valid for other triangular factorisation algorithms. We now write (32) as

$$b_i = (1 + \phi_{ii})^{-1} (1 + \nu_i)^{-1} \ell_{ii} y_i + \sum_{j=1}^{i-1} \ell_{ij} y_j (1 + \theta_{ij})(1 + \phi_{ij})(1 + \phi_{ii})^{-1},$$

which, on neglecting terms of order 2^{-2t}, becomes

$$b_i = \sum_{j=1}^{i} \ell_{ij} y_j (1 + \alpha_{ij}), \qquad \text{for } i = 1, 2, \ldots, n, \tag{33}$$

where $|\alpha_{ii}| \leqslant 2.2^{-t}$ and $|\alpha_{ij}| \leqslant (i - j + 3)2^{-t}$ for $j < i$. Equations (33) show that y is the exact solution of

$$(L + \delta L)\mathbf{y} = \mathbf{b} \tag{34}$$

where bounds for the elements of the perturbation matrix δL are,

$$|\delta \ell_{ij}| = |\alpha_{ij} \ell_{ij}| \leqslant (i - j + 3)2^{-t} \qquad \text{for } j \leqslant i \tag{35}$$

(Here we have used the fact that $|\ell_{ij}| \leqslant 1$ and have neglected the difference between the derived bounds for α_{ii} and α_{ij}.) Dealing similarly with the second of the sets of equations (31):

$$x_i = \left\{ (1 + \phi_{ii}) y_i - \sum_{j=i+1}^{n} u_{ij} x_j (1 + \theta_{ij})(1 + \phi_{ij}) \right\} (1 + \nu_i) \,/\, u_{ii} \tag{36}$$

from which we obtain

$$x_i = \sum_{j=i}^{n} u_{ij} x_j (1 + \beta_{ij}), \qquad \text{for } i = 1, 2, \ldots, n, \tag{37}$$

where $|\beta_{ii}| \leqslant 2.2^{-t}$ and $|\beta_{ij}| \leqslant (n + 4 - j)2^{-t}$. The computed solution $\mathbf{x}$ is thus the exact solution of the equation

$$(U + \delta U)\mathbf{x} = \mathbf{y} \tag{38}$$

where $$|\delta u_{ij}| = |\beta_{ij} u_{ij}| \leqslant (n + 4 - j)2^{-t} |u_{ij}|, \qquad \text{for } i \leqslant j. \tag{39}$$

From (34) and (38), $$(L + \delta L)(U + \delta U)\mathbf{x} = \mathbf{b},$$

and hence, replacing LU by $A + E$, we have

$$(A + E + L\delta U + \delta L \,.\, U + \delta L \,.\, \delta U)\mathbf{x} = \mathbf{b}.$$

Neglecting terms of order 2^{-2t} we see that the computed $\mathbf{x}$ is the exact solution of the perturbed equation

$$(A + \delta A)\mathbf{x} = \mathbf{b} \tag{40}$$

where

$$\delta A = E + L\delta U + \delta L \,.\, U \tag{41}$$

The elements of δA are given by

$$\left.\begin{aligned} \delta a_{ij} &= \epsilon_{ij} + \sum_{k=1}^{i} \{\ell_{ik}\, \delta u_{kj} + \delta \ell_{ik}\, u_{kj}\}, \qquad \text{for } i \leqslant j \\ \delta a_{ij} &= \epsilon_{ij} + \sum_{k=1}^{j} \{\ell_{ik}\, \delta u_{kj} + \delta \ell_{ik}\, u_{kj}\}, \qquad \text{for } i > j \end{aligned}\right\} \tag{42}$$

We have already seen that the strict bounds for elimination with partial pivoting are unacceptable even without rounding errors in the solution of the triangular sets of equations, and so we derive only the bounds based on g, the maximum of the $|a_{ij}^{(k)}|$. Using the bounds (27) for the ϵ_{ij}, and those given in (35) and (39) for $\delta \ell_{ij}$ and δu_{ij} respectively, we obtain

$$|\delta a_{ij}| \leqslant 3(i-1)g\,2^{-t} + \sum_{k=1}^{i} \{(n+4-j)2^{-t}g + (i-k+3)2^{-t}g\}$$

$$= g.2^{-t}\{3(i-1) + (n+7+i-j)i - \tfrac{1}{2}i(i+1)\}$$

$$= g.2^{-t}\{\tfrac{1}{2}i^2 + (n+9\tfrac{1}{2}-j)i - 3\}, \qquad \text{for } i \leqslant j,$$

$$|\delta a_{ij}| \leqslant (3j-2)g.2^{-t} + \sum_{k=1}^{j} \{(n+4-j)2^{-t}g + (i-k+3)2^{-t}g\}$$

$$= g.2^{-t}\{3j-2 + (n+7+i-j)j - \tfrac{1}{2}j(j+1)\}$$

$$= g.2^{-t}\{-\tfrac{3}{2}j^2 + (n+9\tfrac{1}{2}+i)j - 2\}, \qquad \text{for } i > j.$$

From this we may calculate the bound

$$\|\delta A\|_\infty \leqslant \tfrac{1}{2}(n^3 + 10n^2 + 5n - 2)g.2^{-t} \tag{43}$$

The bound for the perturbation due to rounding errors in the elimination process was found to be

$$\|E\|_\infty \leqslant (3n^2 - n - 2)g.2^{-t}$$

From these bounds, it would seem that the greatest errors may arise during the solution of the triangular sets of equations. We have, in fact,

$$\|L\delta U + \delta L U\| \leqslant (\tfrac{1}{2}n^3 + \tfrac{7}{2}n^2 + 3n)g.2^{-t}.$$

This would particularly be the case if the $a_{ij}^{(k)}$ grow steadily throughout the elimination. It could be worthwhile using double length arithmetic in the solution of the triangular sets of equations, even if it had not been used during the elimination. Nevertheless, Wilkinson gives good reason for believing that the rounding errors occurring during solution of the triangular sets of equations will not lead to large errors, and that generally it is the rounding errors in the elimination process which contribute most to inaccuracies in the solution.

5.5 Direct Triangulation Methods and Double-length Arithmetic

5.5.1

In the absence of rounding errors, the Doolittle triangulation algorithm is, in effect, identical with Gaussian elimination. Moreover, if the sum of products which occur, such as $\sum_{k=1}^{r-1} \ell_{rk} u_{kj}$, are evaluated and rounded term by term, then the errors in the Doolittle process are identical with those in Gaussian elimination; those in other direct triangulation methods are of similar magnitude. On the other hand, if double length accumulation of such inner products is available, direct triangulation methods may be considerably more accurate than Gaussian elimination. We shall consider the Doolittle process with double-length accumulation; error bounds for other processes are similar.

Without rounding errors, the elements of the upper and lower triangular matrices are given by

$$\left.\begin{aligned} u_{ij} &= a_{ij} - \sum_{k=1}^{i-1} \ell_{ik} u_{kj}, && \text{for } j = i, \ldots, n, \\ \text{and} \quad \ell_{ij} &= \left[a_{ij} - \sum_{k=1}^{j-1} \ell_{ik} u_{kj}\right] \Big/ u_{jj}, && \text{for } i = j+1, \ldots, n. \end{aligned}\right\} \tag{44}$$

Also, $\ell_{ii} = 1$, for all i. We assume that partial pivoting has been carried out, and that in (44) the a_{ij} are elements of the matrix of coefficients with rows arranged appropriately (so that $|\ell_{ij}| \leqslant 1$). We assume also that the expressions $(a_{ij} - \Sigma \ell_{ik} u_{kj})$ in (44) are calculated as double-length numbers, and in the second case divided by u_{jj}, before being rounded to t-bits. If terms of order 2^{-2t} are ignored, then rounding errors do not more than introduce a single relative error not greater than 2^{-t}, and we can write

$$u_{ij} = (a_{ij} + \nu_{ij} u_{ij}) - \sum_{k=1}^{i-1} \ell_{ik} u_{kj}$$

and

$$\ell_{ij} = \left[(a_{ij} + \nu_{ij} \ell_{ij} u_{jj}) - \sum_{k=1}^{j-1} \ell_{ik} u_{kj}\right] \Big| u_{jj},$$

where $|\nu_{ij}| \leqslant 2^{-t}$.

Thus the matrices U and L are factors of a perturbed matrix of coefficients:

$$LU = A + E \tag{45}$$

$$\left.\begin{aligned} \text{where} \quad |\epsilon_{ij}| &= |\nu_{ij} u_{ij}| \leqslant g 2^{-t}, && \text{for } j = i, \ldots, n \\ \text{and} \quad |\epsilon_{ij}| &= |\ell_{ij} u_{jj} \nu_{ij}| \leqslant g 2^{-t}, && \text{for } i = 1, \ldots, n. \end{aligned}\right. \tag{46}$$

Apart from differences in rounding errors, the u_{ij} are the same as those obtained by Gaussian elimination, and we have again taken $|u_{ij}| \leqslant g$. We have now

$$\|E\|_\infty \leqslant n g 2^{-t}, \tag{47}$$

instead of the bound $(3n^2 - n - 2) g 2^{-t}$ obtained using ordinary single-length arithmetic. If double-length accumulation is also used during the solution of the triangular sets of equations then (33) and (37) become

$$\left.\begin{aligned} b_i &= \sum_{j=1}^{i-1} \ell_{ij} y_j + (1+\alpha_i)\ell_{ii} y_i, \\ \text{and} \qquad y_i &= \sum_{j=i+1}^{n} u_{ij} x_j + (1+\beta_i) u_{ii} x_i, \end{aligned}\right\} \tag{48}$$

so that δL and δU in (41) are now diagonal matrices with bounds

$$\left.\begin{aligned} |\delta \ell_{ii}| &= |\alpha_i \ell_{ii}| \leqslant 2^{-t} \\ \text{and} \quad |\delta u_{ii}| &= |\beta u_{ii}| \leqslant 2^{-t} g \end{aligned}\right\} \tag{49}$$

Using (46) we have bounds for the elements of the total perturbation matrix,

$$\delta A = E + L\delta U + \delta L.U$$

given by

$$\left.\begin{aligned} |\delta a_{ij}| &\leqslant 2g.2^{-t} \quad \text{for } i \neq j \\ |\delta a_{ii}| &\leqslant 3g.2^{-t}. \end{aligned}\right\} \tag{50}$$

and

Hence,

$$\|\delta A\|_\infty \leqslant (2n+1)g.2^{-t}. \tag{51}$$

For any substantial number of equations, this bound is very much smaller than the corresponding bound derived on the assumption that ordinary single-length arithmetic was used.

5.6 Equilibration and Pivoting

5.6.1

In the previous section, we have shown the necessity for limiting the size of the multiplier, m_{ij}, in deriving bounds for the elements of the perturbation matrix, δA. This does not prove that failure to use some sort of pivoting can cause large error, but a simple example, will show that this is the case. The basic rounding errors occur in the calculations: $a_{ij}^{(k+1)} = a_{ij}^{(k)} - m_{i,k+1} a_{k+1,j}^{(k)}$. If $m_{i,k+1} a_{k+1,j}^{(k)}$ is very large compared with $a_{ij}^{(k)}$, then $a_{ij}^{(k)}$ may have no influence on the result. To be specific, if

$$|m_{i,k+1} a_{k+1,j}^{(k)}| > 2^t |a_{ij}^{(k)}|, \tag{52}$$

then the computed value of $a_{ij}^{(k+1)}$, (using a t-bit) mantissa), is $m_{i,k+1} a_{k+1,j}^{(k)}$, and is thus entirely independent of $a_{ij}^{(k)}$. As an example, consider the matrix of coefficients

$$A = \begin{pmatrix} 2^{-t-2} & 1 \\ 1 & 1 \end{pmatrix} \tag{53}$$

Exact Gaussian elimination yields

$$A = \begin{pmatrix} 1 & 0 \\ 2^{t+2} & 1 \end{pmatrix} \begin{pmatrix} 2^{-t-2} & 1 \\ 0 & 1 - 2^{t+2} \end{pmatrix} \tag{54}$$

but with t-bit arithmetic $(1 - 2^{t+2})$ is rounded to -2^{t+2}, and the computed reduction is

$$LU = \begin{pmatrix} 1 & 0 \\ 2^{t+2} & 1 \end{pmatrix} \begin{pmatrix} 2^{-t-2} & 1 \\ 0 & -2^{t+2} \end{pmatrix} = \begin{pmatrix} 2^{-t-2} & 1 \\ 1 & 0 \end{pmatrix} \tag{55}$$

The perturbation matrix due to the rounding error is

$$E = \begin{pmatrix} 0 & 0 \\ 0 & -1 \end{pmatrix},$$

which is of approximately the same norm as is A itself. With specific right-hand sides, the equations

$$\left.\begin{aligned} 2^{-t-2}x + y &= 1 \\ x + y &= 0 \end{aligned}\right\} \tag{56}$$

have the exact solution

$$x = -1/(1 - 2^{-t-2}), \qquad y = 1/(1 - 2^{-t-2}).$$

Using 2^{-t-2} as pivot and rounding to t bits we obtain

$$\left.\begin{aligned} 2^{-t-2}x + y &= 1 \\ -2^{t+2}y &= -2^{t+2} \end{aligned}\right\}$$

from which we find $y = 1$ (which is correct to working accuracy), but $x = 0$ (which is grossly in error). However, if we use partial pivoting, the coefficient of x in the second equation becomes the pivot, and we obtain

$$\begin{aligned} x + y &= 0 \\ y &= 1 \end{aligned}$$

and hence $y = 1, x = -1$. Both unknowns are correct to full t-bit accuracy.

5.6.2

However, partial pivoting is not as yet a complete answer. Consider the equations obtained by multiplying the first equation of (56) by 2^{t+3}:

$$\begin{aligned} 2x + 2^{t+3}y &= 2^{t+3} \\ x + y \quad &= 0 \end{aligned} \tag{57}$$

Partial pivoting would entail choosing the coefficient of x in the first equation as pivot, and would lead to the inaccurate solution $y = 1, x = 0$. To avoid this situation the idea of *equilibration* is introduced. With partial pivoting it is recommended that the equations should be row-equilibrated, i.e. the equations should be scaled so that all the rows of coefficients are of approximately the same magnitude. We may, for example, arrange that the maximum element in each row has modulus lying between $\frac{1}{2}$ and 1. Equations (57) would then be scaled to give again equations (56) which, as we have seen, yield satisfactorily to partial pivoting. It is a matter of experience that partial pivoting with row equilibration is generally satisfactory, and it is probably the most commonly used procedure with Gaussian elimination and triangulation methods. However, we shall carry our examination of equations (57) a little further.

We scale the variables (columns) in equations (57) by writing $z = 2^{t+3}y$. Then we have

$$\begin{aligned} 2x + z &= 2^{t+3} \\ x + 2^{-t-3}z &= 0 \end{aligned} \qquad (58)$$

The exact solution is, of course,

$$x = -1/(1 - 2^{-t-2}), \qquad z = 2^{t+3}/(1 - 2^{-t-2}).$$

The computed solution using partial pivoting is $x = 0, z = 2^{t+3}$. Whether or not this approximate solution is acceptable depends upon the circumstances. Certainly the norm of the computed solution is in error by less than one part in 2^t, but on the other hand if a knowledge of the value of x itself is important, the approximate solution is valueless. Using complete pivoting with equations (58) as they stand will make no difference to this result. If complete pivoting is used with equilibration of rows and columns, the result depends on exactly how the equilibration is carried out. If we divide the first row by 2 and then multiply the second column by 2 we obtain

$$\left.\begin{aligned} x + w &= 2^{t+2} \\ x + 2^{-t-2}w &= 0 \end{aligned}\right\}$$

where $w = 2^{-1}z$. Complete pivoting still leaves the top left-hand coefficient as pivot, and still leads to a computed value of 0 for x.

We repeat, then, Dr. Wilkinson's view that partial pivoting with row equilibration is usually satisfactory, but add the warning that it is possible for it to lead to unsatisfactory results in some (extreme) cases. In equations (58) the single very large term on the right-hand side would give rise to suspicion that such an unsatisfactory result might arise, as would the large discrepancy in size of the values of the unknowns.

5.7 Iterative Improvement of Solutions

5.7.1

Note that none of the sets of equations studied in the previous section are ill-conditioned, and a satisfactory result may be obtained by using the iterative improvement technique described in 3.5.2, even if the "wrong" pivot has been chosen in the triangulation. Consider again equations (56):

$$\left.\begin{aligned} 2^{-t-2}x + y &= 1 \\ x + y &= 0. \end{aligned}\right\}$$

Without partial pivoting the computed triangular decomposition of the matrix of coefficients is

$$A_* = LU = \begin{pmatrix} 2^{-t-2} & 1 \\ 1 & 0 \end{pmatrix}.$$

The inverse of A_* is

$$A_*^{-1} = \begin{pmatrix} 0 & 1 \\ 1 & -2^{-t-2} \end{pmatrix}.$$

The solution is then computed as

$$\mathbf{x}_0 = \begin{pmatrix} x \\ y \end{pmatrix} = \begin{pmatrix} 0 & 1 \\ 1 & -2^{-t-2} \end{pmatrix} \begin{pmatrix} 1 \\ 0 \end{pmatrix} = \begin{pmatrix} 0 \\ 1 \end{pmatrix}$$

as obtained previously. If we now use this triangular decomposition to improve this solution, we compute first

$$\mathbf{r}_0 = \mathbf{b} - A\mathbf{x}_0 = \begin{pmatrix} 1 \\ 0 \end{pmatrix} - \begin{pmatrix} 2^{-t-2} & 1 \\ 1 & 1 \end{pmatrix} \begin{pmatrix} 0 \\ 1 \end{pmatrix} = \begin{pmatrix} 0 \\ -1 \end{pmatrix}$$

and then

$$\mathbf{x}_1 = \mathbf{x}_0 + A_*^{-1}\mathbf{r}_0 = \begin{pmatrix} 0 \\ 1 \end{pmatrix} + \begin{pmatrix} 0 & 1 \\ 1 & -2^{-t-2} \end{pmatrix} \begin{pmatrix} 0 \\ -1 \end{pmatrix} = \begin{pmatrix} -1 \\ 1 \end{pmatrix}$$

where the final vector has been rounded to t-bits. This solution is correct to working accuracy. If we repeat the process we find

$$\mathbf{r}_1 = \mathbf{b} - A\mathbf{x}_1 = \begin{pmatrix} 2^{-t-2} \\ 0 \end{pmatrix}$$

$$\mathbf{x}_2 = \mathbf{x}_1 + A_*^{-1}\mathbf{r}_1 = \begin{pmatrix} -1 \\ 1 \end{pmatrix}$$

(Again the final value of $\mathbf{x}$ has been rounded to t-bits, and $\mathbf{b} - A\mathbf{x}_1$ has been calculated double-length and then rounded to t-bits.) Note that although the approximate inverse A_*^{-1} (equivalent to using the computed triangular decomposition LU) is unsatisfactory when used only once, when used iteratively it yields a result correct to full working accuracy.

CHAPTER 6

COMPUTATION OF EIGENVALUES AND EIGENVECTORS

6.0 Introduction

The aim of this chapter is to introduce some of the most efficient methods of computing eigenvalues and eigenvectors of a real matrix. (For a comprehensive survey of all possible methods see reference [1].) A distinction is made between methods suitable for a real symmetric matrix and those which apply more generally. The special properties of symmetric matrices lead to a reduction in the amount of computer storage needed and also in computing time.

All the methods described here involve three essential stages:

(i) replacement of the original matrix by a similar matrix of some standardised reduced form,
(ii) computation of the eigenvalues of the reduced matrix,
(iii) calculation of eigenvectors, once the eigenvalues are known.

6.1 The Symmetric Eigenvalue Problem

We begin by discussing methods which can be used to find eigenvalues and eigenvectors of a real symmetric matrix. With slight modifications, the same methods can be applied to complex Hermitian matrices, which also have real eigenvalues. In reducing Hermitian matrices, we replace orthogonal transformations ($P' = P^{-1}$) by unitary transformations ($\overline{U}' = U^{-1}$).

6.1.1 Reduction to Tridiagonal Form

Most methods of finding eigenvalues require the evaluation of the characteristic polynomial for a large number of different values of λ. For a *tridiagonal* $n \times n$ matrix evaluation of the characteristic polynomial involves $2n$ multiplications, compared with $n^3/3$ in the general case. Accordingly, given a symmetric matrix A, the first step is to find a similarity transformation reducing A to tridiagonal form. To preserve symmetry during the reduction, we use only orthogonal matrices, i.e. we need a matrix P such that $PP' = I$ and $PAP' = T$, where T is a tridiagonal matrix which has, of course, the same eigenvalues as A.

The orthogonal matrices used in this reduction are usually of one of two types:

(i) Plane rotation matrices.
(ii) Householder's orthogonal transformation matrices. The use of Householder's transformations is generally preferred since, although the individual matrices are more complicated the reduction can be performed in a much smaller number of steps.

6.1.2 Plane Rotation Matrices

In two dimensional coordinate geometry, the orthogonal matrix

$$R = \begin{pmatrix} \cos\theta & -\sin\theta \\ \sin\theta & \cos\theta \end{pmatrix}$$

represents a rotation of the coordinate axes through an angle θ, the new coordinate vector $\mathbf{X}$ being related to the original coordinate vector $\mathbf{x}$ by the equation $\mathbf{X} = R\mathbf{x}$. As a slight extension of this idea, the matrix

$$\begin{pmatrix} \cos\theta & 0 & -\sin\theta \\ 0 & 1 & 0 \\ \sin\theta & 0 & \cos\theta \end{pmatrix}$$

represents in three dimensions a rotation through an angle θ in the plane of the 1st and 3rd coordinate axes. More generally still, an $n \times n$ matrix $R(k, \ell, \theta)$ which differs from the $n \times n$ identity matrix in only four elements, $r_{kk} = r_{\ell\ell} = \cos\theta$, $r_{k\ell} = -r_{\ell k} = -\sin\theta$, is said to represent a rotation through θ in the plane of the kth and ℓth variables. It is left as an exercise for the reader to verify that $R(k, \ell, \theta)$ is an orthogonal matrix and that for any matrix A the similar matrix $R(k, \ell, \theta)AR'(k, \ell, \theta)$ differs from A only in its kth and ℓth rows and columns. A rotation matrix has only one free parameter θ which may be determined from the coefficients of A. It is thus only possible to reduce the non-zero elements of A to zero one at a time by the use of plane rotations. For details of this application of plane rotations see p. 156, Exercise 1.

6.1.3 Householder's Orthogonal Transformation Matrices

Let $\mathbf{u}$ be a real n-vector of unit length, i.e. $\mathbf{u}'\mathbf{u} = \mathbf{u}.\mathbf{u} = u_1^2 + u_2^2 + \ldots + u_n^2 = 1$. Consider the $n \times n$ matrix $P = I - 2\mathbf{u}\mathbf{u}'$.

Then
$$\begin{aligned} PP' &= (I - 2\mathbf{u}\mathbf{u}')(I - 2\mathbf{u}\mathbf{u}') \\ &= I - 4\mathbf{u}\mathbf{u}' + 4\mathbf{u}\,\mathbf{u}'\mathbf{u}\,\mathbf{u} \\ &= I - 4\mathbf{u}\,\mathbf{u}' + 4\mathbf{u}(\mathbf{u}'\mathbf{u})\mathbf{u} \\ &= I, \text{ since } \mathbf{u}'\mathbf{u} = 1. \end{aligned}$$

Hence P is an orthogonal matrix which is, clearly, also symmetric. Rather more importantly, P is completely determined by the n elements of the unit vector $\mathbf{u}$. Unlike the plane rotations, the Householder matrices have $n - 1$ free parameters and, as will be seen, offer the possibility of simultaneously reducing several elements of A to zero. In the next section, the choice of $\mathbf{u}$ and the computation of PAP' will be considered in more detail. Given any non-zero vector $\mathbf{v}$ we shall first show how to find a Householder matrix P such that the vector $\mathbf{w} = P\mathbf{v}$ has all its elements w_i equal to zero except the first.

First suppose that P is of the form
$$P = I - 2\mathbf{u}\mathbf{u}' = I - \frac{1}{k}\mathbf{x}\mathbf{x}'$$

where $\mathbf{u}$ is a unit vector and $\mathbf{x}$ is some multiple of $\mathbf{u}$. Then,
$$\mathbf{w} = P\mathbf{v} = \mathbf{v} - \frac{1}{k}\mathbf{x}\mathbf{x}'\,\mathbf{v}$$

and so,
$$w_1 = v_1 - \frac{1}{k}(\mathbf{x}.\mathbf{v})x_1 \tag{1}$$

while
$$v_i = \frac{1}{k}(\mathbf{x}.\mathbf{v})x_i = 0 \qquad \text{for } i = 2, \ldots, n. \tag{2}$$

Equation (2) will be satisfied if we choose
$$x_i = v_i \qquad \text{for } i = 2, \ldots, n$$
$$\text{and} \quad k = \mathbf{x}.\mathbf{v} \tag{3}$$

It remains to determine x_1. Using (3) and the fact that $\mathbf{u} = \frac{1}{\sqrt{2k}}\mathbf{x}$ must be a unit vector we have
$$k = x_1 v_1 + \sum_{i=2}^{n} v_i^2 \tag{4}$$

$$\text{and} \qquad 2k = \mathbf{x}.\mathbf{x} = x_1^2 + \sum_{i=2}^{n} v_i^2. \tag{5}$$

Solving these last two equations for x_1 we obtain

$$x_1 = v_1 + \pm\sqrt{(v_1^2 + v_2^2 + \ldots + v_n^2)} = v_1 \pm \ell.$$

The sign ambiguity can be removed by adopting the convention

$$x_1 = v_1 + (\text{sign } v_1)\ell.$$

From equation (4) we get,

$$\begin{aligned} k &= x_1 v_1 + \sum_{i=2}^{n} v_i^2 \\ &= \ell^2 + \ell v_1 \\ &= \ell^2 + |\ell v_1|, \text{ using the above convention.} \end{aligned}$$

Full information about the matrix P is retained by storing the constant $k = \ell^2 + |\ell v_1|$ and the vector $\mathbf{x} = (v_1 + (\text{sign } v_1)\ell, v_2, \ldots, v_n)'$. $P\mathbf{v} = \mathbf{w}$ has the very simple form:

$$P\mathbf{v} = \mathbf{w} = (v_1 - x_1, 0, \ldots, 0)' = (\mp\ell, 0, \ldots, 0)'.$$

Note that since P is an orthogonal transformation $\mathbf{v}$ and $\mathbf{w}$ are both vectors of length ℓ.

For any vector $\mathbf{y}$, $P\mathbf{y}$ is quickly computed since

$$\begin{aligned} P\mathbf{y} &= \left(I - \frac{1}{k}\mathbf{x}\mathbf{x}'\right)\mathbf{y} \\ &= \mathbf{y} - \frac{1}{k}(\mathbf{x}.\mathbf{y})\mathbf{x}. \end{aligned}$$

i.e. $P\mathbf{y}$ is obtained from $\mathbf{y}$ by simply subtracting a scalar multiple of $\mathbf{x}$. The matrix product PA is computed in a similar manner by considering the individual products $P\mathbf{c}_i$ where $\mathbf{c}_i, i = 1, \ldots n$ are the column vectors of A. If $B = PA$ then $PAP = BP$ can be computed by noting that $(BP)' = P'B' = PB'$. Hence the row vectors of BP (which are the columns of $(BP)'$) are similarly obtained from the rows of B. When A is a symmetric matrix only the upper (or lower) triangle of PAP need be computed.

6.1.4 Householder's Method of Reducing A to Tridiagonal Form

Suppose A is a real symmetric $n \times n$ matrix. We wish to find an orthogonal matrix P such that

$$P'AP = \begin{pmatrix} d_1 & c_1 & 0 \dots & & 0 \\ c_1 & d_2 & c_2 & \dots & \\ 0 & c_2 & d_3 & \dots & \\ \vdots & & & & \\ 0 & \dots & c_{n-2} & d_{n-1} & c_{n-1} \\ 0 & \dots & & c_{n-1} & d_n \end{pmatrix}$$

This reduction can be performed as a product of Householder transformations, $P = P_{n-2}P_{n-3} \dots P_2P_1$. As a first stage of the reduction we require a matrix P_1 such that

$$P_1AP_1 = \left(\begin{array}{c|ccc} d_1 & c_1 & 0 \dots & 0 \\ \hline c_1 & & & \\ 0 & & A_1 & \\ \vdots & & & \\ 0 & & & \end{array}\right)$$

where A_1 is a symmetric matrix of order $n-1$.

If P_1 is partitioned in the form $P_1 = \left(\begin{array}{c|c} 1 & 0 \\ \hline 0 & Q_1 \end{array}\right)$, where Q_1 is a Householder matrix of order $n-1$, we obtain:

$$\begin{aligned} P_1AP_1 &= \left(\begin{array}{c|c} 1 & 0 \\ \hline 0 & Q_1 \end{array}\right) \left(\begin{array}{c|c} a_{11} & \mathbf{z}' \\ \hline \mathbf{z} & B_1 \end{array}\right) \left(\begin{array}{c|c} 1 & 0 \\ \hline 0 & Q_1 \end{array}\right) \\ &= \left(\begin{array}{c|c} a_{11} & \mathbf{z}' \\ \hline Q_1\mathbf{z} & Q_1B_1 \end{array}\right) \left(\begin{array}{c|c} 1 & 0 \\ \hline 0 & Q_1 \end{array}\right) = \left(\begin{array}{c|c} a_{11} & (Q_1\mathbf{z})' \\ \hline Q_1\mathbf{z} & Q_1B_1Q_1 \end{array}\right) \end{aligned}$$

This product matrix will then be of the required form if Q_1 is the Householder matrix which annihilates all elements of the vector $\mathbf{z}$ except the first. Q_1 is thus computed in the manner described in the previous paragraphs from the sub-diagonal elements in the first column of A.

From this computation we obtain $d_1 = a_{11}$ and $c_1 = \mp \ell$, where ℓ is the length of the first subdiagonal vector of A and the sign is opposite to that of the first element of this vector. Similarly at the rth stage of the reduction

$$P_r = \left(\begin{array}{c|c} I_r & 0 \\ \hline 0 & Q_r \end{array}\right)$$

where I_r is the rth order identity matrix and Q_r is the $(n-r) \times (n-r)$

Householder matrix which annihilates all but the leading element of the rth subdiagonal column vector of the matrix currently being reduced. Note that pre- and post-multiplying by P_r has no effect upon the 1st $(r-1)$ columns and rows which have previously been reduced to tridiagonal form.

Computational Details

Since the ultimate aim is to find the eigenvalues and eigenvectors of the original matrix A, it is essential to store not only the elements of the tridiagonal matrix to which A is reduced but also the details of the orthogonal transformations used in this reduction. A convenient way of arranging the storage of this information is as follows:

(i) Retain the elements $d_1, d_2, \ldots, d_n$ and $c_1, \ldots, c_{n-1}$, as vectors.
(ii) Overwrite the matrix A as it is reduced, but store in the subdiagonal positions the vectors $\mathbf{x}_1, \mathbf{x}_2, \ldots$, which define the Householder transformations $Q_i = I - \frac{1}{k_i}\mathbf{x}_i\mathbf{x}_i'$ used in the reduction.

[This requires very little modification of the matrix since all but the first element of $\mathbf{x}_i$ will be equal to the corresponding elements of the column vector currently in this position.]

The constants k_i can be stored in the diagonal positions as they are computed.

Alternatively, since for general P we have $k = \ell^2 + |\ell v_1|$ and $x_1 = v_1 + (\text{sign } v_1)\ell$, then with $c_i = \mp\ell_i$ we obtain $k_i = -c_i x_1$. This enables us to calculate the constant k_i from the subdiagonal element c_i and the first element x_1 of the vector x_i. Using this method, $d_1, d_2, \ldots$ can be stored as diagonal elements of the matrix as it is reduced, the constants $k_1, k_2, \ldots$ being calculated when needed.

Since this method of reduction depends upon the use of orthogonal transformations it is basically a numerically stable process (see Chapter 7). The accuracy of the computed elements of the tridiagonal matrix can be improved at the cost of a small increase in computer time if all scalar products are computed using double precision arithmetic.

6.1.5 Given's Method of Reducing A to Tridiagonal Form

Given's method makes use of plane rotations to systematically reduce a symmetric $n \times n$ matrix to tridiagonal form. Using this method it is only possible to annihilate the subdiagonal elements of A one at a time, so that it requires $\frac{1}{2}(n-1)(n-2)$ such rotations to perform the complete reduction. This is to be compared with the corresponding reduction using $(n-2)$ Householder transformations. Allowing for the fact that the individual plane rotations are simpler than the Householder transformations, it takes approximately twice as many arithmetic operations to reduce a matrix to tridiagonal form by Given's method as by

Householder's method. The elements are annihilated in the order (3, 1)(4, 1), . . . $(n, 1), (4, 2), \ldots (n, 2), \ldots (n, n-2)$, using a rotation in the $(i+1, j)$ plane to annihilate the element a_{ij}.

6.1.6 Eigenvalues of a Tridiagonal Symmetric Matrix

Once the symmetric matrix A has been reduced to tridiagonal form, the next problem is that of finding the eigenvalues of this reduced matrix. The method described below depends upon using the *Sturm sequence property* of the characteristic polynomial, and was first used by Givens.

Let T be a real, symmetric, tridiagonal matrix whose diagonal elements are $d_1, d_2, \ldots, d_n$ and whose subdiagonal elements are $c_1, c_2, \ldots, c_{n-1}$. The characteristic polynomial of T is then

$$f_n(\lambda) = \begin{vmatrix} d_1-\lambda & c_1 & 0 & \cdots & 0 \\ c_1 & d_2-\lambda & c_2 & & \\ 0 & & & & \\ & & & & 0 \\ & & & & c_{n-1} \\ 0 & \cdots & 0 & c_{n-1} & d_n-\lambda \end{vmatrix}$$

If we now define $f_r(\lambda)$ to be the determinant obtained from this by removing the last $n-r$ rows and columns we have:

$$f_1(\lambda) = d_1 - \lambda, \quad f_2(\lambda) = \begin{vmatrix} d_1-\lambda & c_1 \\ c_1 & d_2-\lambda \end{vmatrix} = (d_2-\lambda)(d_1-\lambda) - c_1^2$$

and so on.

Expanding $f_n(\lambda)$ by its last row we obtain:

$$f_n(\lambda) = (d_n - \lambda) f_{n-1}(\lambda) - c_{n-1}^2 f_{n-2}(\lambda).$$

More generally, $$f_m(\lambda) = (d_m - \lambda) f_{m-1}(\lambda) - c_{m-1}^2 f_{m-2}(\lambda). \quad (6)$$

Equation (6) is true for $m = 2, \ldots, n$ if we formally define $f_0(\lambda) = 1$. We can assume that none of the subdiagonal elements $c_1, c_2, \ldots, c_{n-1}$ is zero. [Otherwise if $c_i = 0$ then $f_i(\lambda)$ is a factor of $f_n(\lambda)$ and the order of the problem is immediately reduced.]

We shall prove that the sequence of polynomials $f_0(\lambda), f_1(\lambda), \ldots, f_n(\lambda)$, has the so-called Sturm Sequence property:

Theorem

The number of agreements in sign between consecutive terms in the sequence $f_0(\lambda), f_1(\lambda), \ldots, f_n(\lambda)$ is equal to the number of roots of the characteristic polynomial which are strictly greater than λ.

Proof

From equation (6) we can deduce that if two consecutive polynomials $f_m(\lambda)$, $f_{m-1}(\lambda)$, have a common root then this root must be common to all polynomials in the sequence, including $f_0(\lambda)$. Since $f_0(\lambda) \equiv 1$ it is clearly impossible for consecutive polynomials to have a common root.

Now let X be any positive number which is greater than the modulus of the largest root of any of the polynomials. Consider the effect upon the number of sign agreements in the sequence of polynomials as λ steadily increases from $-X$ to X. It is clear from the definitions of the $f_m(\lambda)$ that $f_0(-X), f_1(-X), \ldots, f_n(-X)$, are all positive; hence there are n sign agreements in this case. Again, the numbers $f_0(X), f_1(X), \ldots, f_n(X)$, alternate in sign and so there are no sign agreements. Further, if λ' is a root of one of the intermediate polynomials, $f_{m-1}(\lambda)$, in the sequence, we must have

$$f_m(\lambda') = (d_m - \lambda')f_{m-1}(\lambda') - c_{m-1}^2 f_{m-2}(\lambda) = -c_{m-1}^2 f_{m-2}(\lambda').$$

Thus $f_m(\lambda')$ and $f_{m-2}(\lambda')$ must differ in sign. Hence, as λ increases through the value λ', the number of sign agreements in the sequence

$$f_{m-2}(\lambda), \quad f_{m-1}(\lambda), \quad f_m(\lambda),$$

is unaltered. But, as λ increases through a root λ_i of the characteristic polynomial $f_n(\lambda)$, the number of sign agreements in the entire sequence is altered, since, for small ϵ, $f_n(\lambda_i - \epsilon)$ and $f_n(\lambda_i + \epsilon)$ must differ in sign, and only one can agree with the sign of $f_{n-1}(\lambda_i)$.

It follows that the number of sign agreements is decreased by 1 whenever λ increases through an eigenvalue of T, and hence that the number of eigenvalues greater than some given λ_0 is equal to the number of sign agreements in $f_0(\lambda_0)$, $f_1(\lambda_0), \ldots, f_n(\lambda_0)$.

Note. We can also deduce that the roots of $f_{(n-1)}(\lambda)$ lie between those of $f_n(\lambda)$, but this fact is not used in the computational algorithm. From this, and the earlier result that consecutive polynomials in the sequence cannot have common roots, it follows that a tridiagonal symmetric matrix cannot have repeated roots unless there are zeros among the subdiagonal elements. In theory, this should mean that whenever repeated roots are present the order of the problem can be reduced. However, in practice rounding errors can introduce non-zero elements when exact arithmetic would give zero elements; an efficient algorithm should therefore make provision for the possibility of repeated roots occurring.

A Practical Algorithm

The Sturm Sequence property is the basis of a practical algorithm for computing the eigenvalues of a tridiagonal symmetric matrix to a high degree of accuracy. The simplest approach is to use a bisection method after first determining absolute bounds for the eigenvalues.

Let $$X = \max_{i=1,\ldots,n} \{|c_i| + |d_i| + |c_{i-1}|\}.$$

Then, since $X = \|T\|_1$, all the eigenvalues of T lie in the range $-X \leqslant \lambda \leqslant X$. Initially, we can set $a = -X$, $b = X$, and be confident that all the eigenvalues $\lambda_1 > \lambda_2 > \ldots > \lambda_n$ are contained in the interval $[a, b]$. This interval can then be reduced in length by the bisection method described below until it is as small as we wish and contains just one specific eigenvalue, λ_k.

The algorithm which achieves this bisection can be described as follows:

Let $x = \frac{1}{2}(a + b)$
and m = number of sign agreements in $f_0(x), f_1(x), \ldots, f_n(x)$.
Then, if $m < k$ we set $b = x$; otherwise we set $a = x$ (since we know that if $m < k$ there are less than k eigenvalues greater than x).

This bisection process can be continued until the interval (a, b) is shorter than the predetermined accuracy required for λ_k, or until (a, b) is of a length comparable with the relative precision of the computer (in which case $x = \frac{1}{2}(a + b)$ will coincide either with a or with b).

Once λ_k has been determined we can go on to find λ_{k-1} or λ_{k+1} by a repetition of the bisection process starting either with $a = \lambda_k$, $b = X$, or with $a = -X$, $b = \lambda_k$, as appropriate. The bisection method dsecribed has the merit of simplicity, but a more sophisticated criterion for determining x can considerably reduce the number of evaluations of $f_i(\lambda)$ needed. (See reference [4].)

In practice, the Givens–Householder method is frequently used when only a predetermined number of the smallest, or largest, eigenvalues are required. If all the eigenvalues are required, it may be preferable, for a large-matrix, to use the Q–R algorithm described later.

6.1.7 Calculation of the Eigenvectors

Once the eigenvalues of the reduced (tridiagonal) matrix have been obtained, there remains the problem of finding the corresponding eigenvectors.

In theory, it would appear that all that is necessary is to solve, for any given λ_i, the equation:

$$(A - \lambda_i I)\mathbf{x}_i = 0. \tag{7}$$

In practice, λ_i is subject to computing errors, and the matrix $A - \lambda_i I$ may be non-singular, in which case the only solution of (7) is $\mathbf{x}_i = \underline{0}$. A more practical approach to the problem is to use the method of inverse iteration to find the eigenvector of the tridiagonal matrix T and then convert this to an eigenvector of A by using information retained about the orthogonal reduction process.

6.1.8 Method of Inverse Iteration

Assume $\bar{\lambda}$ is a computed approximation to the eigenvalue λ_i of the tridiagonal matrix T.

Unless $\bar{\lambda}$ is precisely equal to λ_i, $T-\bar{\lambda}I$ will be a non-singular matrix and $(T-\bar{\lambda}I)^{-1}$ has the eigenvector $\mathbf{x}_i$ associated with the large eigenvalue $(\lambda_i-\bar{\lambda})^{-1}$. Using the power method with $B=(T-\bar{\lambda}I)^{-1}$, as iteration matrix should quickly produce a good approximation to $\mathbf{x}_i$; in practice, two iterations are usually sufficient if $\bar{\lambda}$ is a good approximation to λ_i.

The inverse iteration process can be summarized as follows:

$$\text{(i) Solve } (T-\lambda I)\mathbf{z}_1=\mathbf{y}_0, \tag{8}$$

$$\text{(ii) Set } \mathbf{y}_1=\frac{1}{\|\mathbf{z}_1\|}\mathbf{z}_1,$$

$$\text{(iii) Solve } (A-\lambda I)\mathbf{z}_2=\mathbf{y}_1.$$

Then $\mathbf{y}_2=\dfrac{1}{\|\mathbf{z}_2\|}\mathbf{z}_2$ is a good approximation to the required eigenvector $\mathbf{x}_i$ of T provided $\mathbf{y}_0$ contains a non-zero term in $\mathbf{x}_i$ when expanded as a linear combination of the eigenvectors of T. In practice, equation (8) is solved by obtaining an $L.U$ factorisation of $(T-\bar{\lambda}I)$, [particularly simple for a tridiagonal matrix] and then obtaining $\mathbf{z}_1$ from the condition $L\mathbf{z}_1=\mathbf{u}$, where $\mathbf{u}=(1,1,\ldots 1)'$. Inevitably, since $(T-\lambda I)$ is nearly singular, this factorisation will contain one or more small pivots. If a zero pivot occurs, it can be replaced by an arbitrarily small number without invalidating the method.

If λ_i is a repeated eigenvalue, we will require to find the appropriate number of linearly independent eigenvectors. The first eigenvector can be found by the method outlined above, subsequent eigenvectors are found either by replacing the estimate $\bar{\lambda}$ by $\bar{\lambda}+\epsilon$, where ϵ is small, or by replacing $\mathbf{u}$ by any other vector, and then applying the standard method.

Example

$$T=\begin{pmatrix}4 & 5 & 0\\ 5 & -1 & -5\\ 0 & -5 & 4\end{pmatrix}$$

This matrix has exact eigenvalues $\lambda_1=9, \lambda_2=4, \lambda_3=-6$. Taking $\bar{\lambda}=8.99$ as an approximation to λ_1 and working throughout to an accuracy of four significant figures, we obtain

$$(T-\bar{\lambda}I)=\begin{pmatrix}1 & 0 & 0\\ -1.002 & 1 & 0\\ 0 & 1.004 & 1\end{pmatrix}\begin{pmatrix}-4.99 & 5 & 0\\ 0 & -4.98 & -5\\ 0 & 0 & 0.030\end{pmatrix}=LU$$

(Note: We have not attempted a Cholesky factorisation of $T-\bar{\lambda}I$ since, unless $\bar{\lambda}$ is less than the least eigenvalue of T, $T-\bar{\lambda}I$ cannot be positive definite.)

Solving $L\mathbf{z}_1 = (1, 1, 1)'$ gives $\mathbf{z}_1 = (-33.93, -33.66, 33.33)'$.
Taking $\|\mathbf{z}_1\|_\infty$ as the scaling factor we obtain:

$$\mathbf{y}_1 = (1, 0.9923, -0.9825)'$$
$$U\mathbf{x} = \mathbf{y}_1 \text{ gives } \mathbf{x} = (1.0, 1.994, -2.984)'$$
$$L\mathbf{z}_2 = \mathbf{x} \text{ gives } \mathbf{z}_2 = (99.48, 99.48, -99.48)'$$
$$\text{Hence } \mathbf{y}_2 = (1.000, 1.000, -1.000)'$$

This is in fact the exact eigenvector for $\lambda_1 = 9$. It is left as an exercise for the reader to find, in a similar way, the eigenvectors corresponding to $\bar{\lambda}_2 = 4.01$ and $\bar{\lambda}_3 = -5.99$.

The method of inverse iteration is not only applicable to symmetric tridiagonal matrices. It can in fact be used for any matrix for which an approximate eigen value is available; obviously if the matrix is not tridiagonal the LU factorisation will be more complicated. [See Exercise 2, p. 156.]

6.1.9 Other Method for Symmetric Matrices

One of the oldest algorithms for finding all the eigenvalues of a full symmetric matrix is the *Jacobi method.* The classical version of this algorithm, which dates from the 19th century, uses repetitively a plane rotation to eliminate the off-diagonal element which is currently of largest modulus in the matrix A. To eliminate the elements a_{ij} and a_{ji} a rotation in the (i, j) plane is used. This will affect only elements in the ith and jth rows and columns of the matrix. Subsequent rotations will re-introduce non-zero elements where zeroes have been previously created; but with each rotation the sum of the squares of the off-diagonal elements is reduced and ultimately the process converges towards a diagonal matrix. Using a computer, the search at each step for the largest off-diagonal element can be time consuming and an alternative is to eliminate the sub-diagonal elements in some predetermined order. Several such eliminations will be required before the matrix approximates to diagonal form. Lanczos' method [3] provides an alternative method of reducing a symmetric matrix to tridiagonal form.

6.2 The Non-symmetric Eigenvalue Problem

If A is a real, non-symmetric, matrix the problem of finding its eigenvalues and eigenvectors is complicated by the facts that A may have complex eigenvalues and that when A has repeated eigenvalues it may not have a sufficient number of corresponding linearly independent eigenvectors. As with symmetric matrices, the problem is usually tackled in three stages, first reducing A to a simpler form (in this case *upper Hessenberg*), then finding the eigenvalues and eigenvectors of the reduced matrix, and finally recovering from these the eigenvectors of the original matrix. If only the eigenvalues are required the computation can be simplified by omitting the last stage and not retaining any information about the transformations used in the reduction process.

6.2.1 Reduction to A to Upper Hessenberg Form

We shall assume that A is a real $n \times n$ matrix. We require to find a non-singular matrix P such that $P^{-1}AP = H$, where H is a matrix with zeroes throughout the lower triangle, except in the positions adjacent to the diagonal elements.

If we can perform this reduction A and H will have the same eigenvalues, and the eigenvectors of A are obtainable from those of H by pre-multiplying by P. There are several ways of performing this reduction. (See Wilkinson [2].) Two of the best are the Householder method (which has the advantage of using orthogonal transformations which ensure that the eigenproblem of H is as well conditioned as that of A) and the use of elementary transformations (which has the advantage of simplicity).

6.2.2 The Householder Method

If we apply the method outlined in Section 6.1.4 to a real non-symmetric $n \times n$ matrix A, a sequence of $n - 2$ Householder transformations determined by the sub-diagonal elements of the first $n - 2$ columns of A will again have the effect of annihilating all but the leading elements of these columns. However, since A is not symmetric, the corresponding elements in the upper triangle of the reduced matrix will, in general, be non-zero. The overall effect is thus to produce an orthogonally similar matrix $H = PAP'$ which is of upper Hessenberg rather than of tridiagonal form. As with the Givens–Householder process, if the eigenvectors of A will subsequently be required, it is necessary to store the relevant information about the Householder transformation matrices used in this reduction.

6.2.3 The Elementary Transformation Method

The object of this method is to successively reduce a real $n \times n$ matrix A to upper-Hessenberg form by a sequence of similarity transformations given by elementary transformation matrices $E_1, E_2, \ldots, E_{n-2}$:

$$A \to E_1 A E_1^{-1} \to E_2 E_1 A E_1^{-1} E_2^{-1} \to \ldots \to E_{n-2} \ldots E_1 A E_1^{-1} \ldots E_{n-2}^{-1} = H.$$

Unlike the Householder method which uses orthogonal transformation, the numerical stability of the process cannot be guaranteed and the matrix H may not be as well conditioned as the original matrix A. Provided the original matrix A does not have ill-conditioned eigenvalues and vectors, and that partial pivoting is used in the reduction process, this objection to the method is of only theoretical importance. Compared with the Householder method, there is a saving of approximately 50% in the number of arithmetic operations performed in the reduction process.

To see how the reduction is performed, we shall consider in detail the rth step of the process. If A_r denotes the matrix obtained from A by the first $r-1$ similarity transformations, we can assume that the first $r-1$ columns of A are of upper Hessenberg form:

$$A_r = \begin{pmatrix} a_{11} & \dots & a_{1r-1} & a_{1r} & \dots & a_{1n} \\ a_{21} & & a_{2r-1} & a_{2r} & & a_{2n} \\ 0 & & & & & \\ \vdots & & & \vdots & & \vdots \\ & & a_{rr-1} & a_{rr} & & a_{rn} \\ & & 0 & a_{r+1r} & & \\ & & \vdots & \vdots & & \vdots \\ 0 & & 0 & a_{nr} & & a_{nn} \end{pmatrix}$$

$E_rA_rE_r^{-1}$ is required to be of a similar form but with the elements $a_{r+2,r}, \dots, a_{n,r}$, of the rth column annihilated. E_r is defined by a sequence of elementary row operations which perform this annihilation:

(a) Find the largest modulus sub-diagonal element of the rth column of A, interchanging rows if necessary to bring this to position $a_{r+1,r}$ (the pivot interchange).

(b) For $k = r+2, \dots, n$, subtract a_{kr}/a_{r+1r} times row $r+1$ from row k.

Notes: The suffix notation of a_{ij} refers to the element currently in row i column j.

If the pivot element is either zero or negligible, E_r is defined to be the identity matrix I and the computation advances to find E_{r+1}.

The effect of the row operations described is to alter only the last $n-r$ rows of A and to leave intact the zero elements in the first $r-1$ columns. The matrix obtained at this stage is E_rA_r. To compute $E_rA_rE_r^{-1}$ the corresponding inverse operations must be performed on the columns of this matrix. These operations are:

(a) Interchange column $r+1$ with a subsequent column.

(b) For $k = r+2, n$ add a_{kr}/a_{r+1r} times column k to column $r+1$.

As can be seen, only two columns of the matrix are affected by these operations and in particular the zeroes in the first r columns are preserved.

If the eigenvectors of A are required, it is necessary to store information about these transformations. This can be done in an economical way if an integer vector is stored to define the row interchanges, and the multipliers used in the elimination are then over-written on the zeroes in the lower triangle of A (i.e. if, in E_r, row $r + 1$ is interchanged with row s, the integer s is stored in the rth position of the vector, and a_{kr}/a_{r+1r} is stored in position (k, r)).

Once the eigenvectors of H have been computed those of A are obtained by performing the appropriate inverse row operations in reverse order upon these vectors (starting with E_{n-2}^{-1}).

Example

Find an upper Hessenberg matrix which is similar to

$$A = \begin{pmatrix} 1 & 2 & 1 & 3 \\ 1 & 3 & 1 & -1 \\ -1 & 0 & 1 & 2 \\ 2 & 4 & 2 & 0 \end{pmatrix}$$

Using elementary transformations as described above:

$$E_1A = \begin{pmatrix} 1 & 2 & 1 & 3 \\ 2 & 4 & 2 & 0 \\ 0 & 2 & 2 & -1 \\ 0 & 1 & 0 & -1 \end{pmatrix}$$

(This is obtained by interchanging rows 2 and 4, and then subtracting $-\frac{1}{2}$ times row 2 from row 3 and $\frac{1}{2}$ row 2 from row 4.)

Then

$$E_1AE_1^{-1} = \begin{pmatrix} 1 & \frac{7}{2} & 1 & 2 \\ 2 & 1 & 2 & 4 \\ 0 & -1 & 2 & 2 \\ 0 & -\frac{1}{2} & 0 & 1 \end{pmatrix}$$

(Obtained after interchanging the columns and adding the appropriate multiples of the subsequent columns to column 2.)

It is left as an exercise for the reader to show that

$$H = E_2E_1AE_1^{-1}E_2^{-1} = \begin{pmatrix} 1 & \frac{7}{2} & 2 & 2 \\ 2 & 1 & 4 & 4 \\ 0 & -1 & 3 & 2 \\ 0 & 0 & -1 & 0 \end{pmatrix}$$

6.2.4 The *QR* Algorithm

Before describing in detail the specialised form of the $Q-R$ algorithm as applied to upper Hessenberg matrices we shall give a brief description of the basic algorithm.

If A is any real $n \times n$ matrix, we can find an orthogonal matrix Q and an upper triangular (or, right triangular) matrix R such that $A = QR$. In practice, it is easier to find Q' such that $Q'A = R$; Q' can be expressed as a product of Householder matrices which successively annihilate elements in the lower triangle of A. After factorising A, the product RQ defines a matrix A_1 which is similar to A since $RQ = Q'AQ$. Continuing in this way, we obtain an iterative sequence of similar matrices defined by the equations:

$$\begin{aligned} Q_m R_m &= A_m \\ R_m Q_m &= A_{m+1} \end{aligned} \tag{9}$$

It can be proved (see Wilkinson [1]) that, when A has real eigenvalues which are all distinct in modulus, the process ultimately produces a diagonal matrix. Clearly the diagonal elements of $\lim_{m \to \infty} A_m$ are the eigenvalues of A, and these are arranged down the diagonal in order of descending modulus. The rate of convergence is dependent upon the rate at which $\lim \left(\frac{\lambda_i}{\lambda_{i-1}}\right)^m \to 0$, where $|\lambda_1| > |\lambda_2| > \ldots > |\lambda_n|$. If A has complex conjugate eigenvalues, the final matrix will not be triangular but will have real 2×2 submatrices along the diagonal with the appropriate complex conjugate eigenvalues. In this simple form, the QR algorithm has convergence properties which are little better than those of the power method, but when it is combined with origin shifts and applied to an upper Hessenberg matrix it is a very powerful algorithm.

The rate at which the QR algorithm converges to produce λ_i as a diagonal element can be accelerated by introducing an origin shift k so that $\left|\frac{\lambda_i - k}{\lambda_j - k}\right|$ is minimised. To allow for the possibility of using a different shift k_m at each iteration without the necessity of keeping a record of all the shifts used, after obtaining the QR factorisation the shift can be restored. In this modified form equations (9) become:

$$\begin{aligned} A_m - k_m I &= Q_m R_m \\ R_m Q_m + k_m I &= A_{m+1}. \end{aligned} \tag{10}$$

After this restoration A_{m+1} is still a matrix with the same eigenvalues as A_m. The use of origin shifts will, however, spoil the final ordering of the eigenvalues along the diagonal.

6.2.5 Application of the *QR* Algorithm to Upper Hessenberg Matrices

Even with origin shifts to accelerate the convergence the QR algorithm is not suitable for use with full matrices because of the large amount of computation needed to perform the factorisation at each iteration. This in itself is comparable with the reduction of a full matrix to upper Hessenberg form. If the matrix A_m of equation (10) is of upper Hessenberg form then the matrices Q_m and A_{m+1} are also of this form since the product of an upper triangular matrix and an upper Hessenberg matrix is of upper Hessenberg form. Also, the amount of computation in obtaining Q_m is reduced drastically since Q_m is the product of Householder matrices related to vectors with only two non-zero elements (plane rotations). From the above, it can be seen that the QR iterative process is considerably simplified if it starts with an upper Hessenberg matrix.

The usual strategy is to employ origin shifts which are eigenvalues of the 2 x 2 matrix currently in the lower right-hand corner of A_m. The disadvantage of this strategy is that these eigenvalues will on occasion be complex, and the QR factorisation can no longer be performed with real arithmetic. The double shift QR algorithm of Francis overcomes this difficulty by making shifts in consecutive iterations which, when necessary, are complex conjugates, and so using real arithmetic to compute the result of the two iterations.

From equations (10) we have, for two consecutive iterations;

$$A_{m+1} = Q'_m A_m Q_m, \qquad Q_m R_m = A_m - k_m I$$
$$A_{m+2} = Q'_{m+1} A_{m+1} Q_{m+1}, \qquad Q_{m+1} R_{m+1} = A_{m+1} - k_{m+1} I$$

Thus, $A_{m+2} = Q'_{m+1} Q'_m A_m Q_m Q_{m+1} = Q' A_m Q$, where Q denotes $Q_m Q_{m+1}$. If R denotes $R_{m+1} R_m$ then R is upper triangular and:

$$\begin{aligned} QR &= Q_m Q_{m+1} R_{m+1} R_m = Q_m (A_{m+1} - k_{m+1} I) R_m \\ &= Q_m (Q'_m A_m Q_m - k_{m+1} I) R_m \\ &= (A_m - k_{m+1} I) Q_m R_m = (A_m - k_{m+1} I)(A_m - k_m I). \end{aligned}$$

Thus Q and R satisfy:

$$QR = (A_m - k_{m+1} I)(A_m - k_m I) \tag{11}$$

and

$$A_{m+2} = Q' A_m Q \tag{12}$$

Let $B = (A_m - k_{m+1} I)(A_m - k_m I) = A_m^2 - (k_m + k_{m+1}) A_m + k_m k_{m+1} I$. Then if A_m is a real, upper Hessenberg, matrix and k_m, k_{m+1}, are either both real or else complex conjugates, B is a real matrix which will in general have two non-zero subdiagonal elements in each column. Equation (11) can be expressed as $R = Q'B$, and Q' could be computed as the orthogonal matrix which annihilates the subdiagonal elements of B. Q' would then be the product of $n - 1$ Householder transformations, each defined by a vector with three non-zero elements, $Q' = P_{n-2} P_{n-3} \ldots P_1 P_0$. In fact, it is not necessary to compute all of matrix B since, after P_0 has been determined, the remaining transformations are determined by equation (12) which requires that $P_{n-2} \ldots P_1 P_0 P_m P_0 \ldots P_{n-2}$ should be an upper Hessenberg matrix.

The first column vector of B is of the form $(b_{11} b_{21} b_{31} 0 \ldots 0)'$, and P_0 is the Householder matrix which annihilates b_{21} and b_{31}. Thus $P_0 = I - 2\mathbf{u}_0 \mathbf{u}'_0$, where

$\mathbf{u}_0$ has non-zero elements only in the first three positions. If $A_m \equiv \{a_{ij}\}$ then a direct calculation of the elements of the first column of $(A_m - k_{m+1}I)(A_m - k_m I)$ gives:

$$b_{11} = (a_{11} - k_{m+1})(a_{11} - k_m) + a_{12}a_{21},$$

$$b_{21} = a_{21}(a_{11} - k_m) + (a_{22} - k_{m+1})a_{21},$$

$$b_{31} = a_{32}a_{21}.$$

Once P_0 has been determined from the above equations, $P_0 A_m P_0$ can be computed.

$P_0 A_m$ differs from A_m only in the first three rows and is of the form,

$$\left(\begin{array}{cccccc} x & x & x & \cdots & & x \\ x & x & \cdots & & & x \\ x & x & x & & & x \\ \hline 0 & 0 & x & & & x \\ \vdots & \vdots & 0 & & & \vdots \\ 0 & 0 & \vdots & & 0 \quad x & x \end{array}\right)$$

where only elements above the partition differ from those of A_m.

Post-multiplication by P_0 modifies the first three columns of this matrix and the result is:

$$P_0 A_m P_0 = \left(\begin{array}{ccc|cccc} x & x & x & x & \cdots & & x \\ x & x & x & & & & \vdots \\ \textcircled{x} & x & x & & & & \vdots \\ \textcircled{x} & \textcircled{x} & x & & & & \\ 0 & 0 & 0 & x & & & \\ & & & 0 & & & \\ & & & \vdots & & & \\ 0 & & & 0 & \cdots\ 0 & x & x \end{array}\right)$$

where only elements to the left of the partition have been modified by the post-multiplication.

In this matrix, only the ringed elements are not conforming to Hessenberg form. Following the Householder method, the unwanted elements in the first column can be annihilated by a Householder matrix P_1 related to a vector having non-zero elements in the 2nd, 3rd and 4th positions only. $P_1 P_0 A_m P_0 P_1$ differs from the above matrix only in rows 2 to 4 and columns 2 to 4. The overall effect is to move the location of the three 'extra' non-zero elements one step down the diagonal. P_2 is then uniquely defined by the requirement to annihilate the elements in the positions a_{42} and a_{52}. Continuing in this way, the non-zero terms are progressively moved down the diagonal until, after the similarity transformation given by P_{n-2}, the matrix is returned to upper Hessenberg form. Precisely the same sequence of Householder matrices could have been computed from equation (11) but the method described avoids the necessity to compute more than three elements of B.

Once A_{m+2} has been computed in the above way, the subdiagonal elements of this Hessenberg matrix are searched to see whether the problem can be deflated. The action to be taken depends upon the location of zero, or negligibly small, elements in the subdiagonal. If $a_{n,n-1} \simeq 0$ then a_{nn} is a real eigenvalue and the matrix can be deflated by removing the last row and column before proceeding to the next QR double shift iteration. If, instead, $a_{n-1\,n-2} \simeq 0$ then two of the eigenvalues of A are given by solving

$$\begin{vmatrix} a_{n-1\,n-1} - \lambda & a_{n-1\,n} \\ a_{nn-1} & a_{nn} - \lambda \end{vmatrix} = 0.$$

Complex conjugate eigenvalues will always appear in this way, and occasionally pairs of real eigenvalues will be simultaneously produced. The matrix is then deflated by omitting the last two rows and columns before proceeding to the next iteration. If $a_{k,k-1}$ is negligible for $k \neq n$, or $n-1$, no eigenvalues are immediately produced but the matrix can be deflated by omitting the first $k-1$ rows and columns.

{In this case A_{m+2} is approximately of the form:

$$\begin{pmatrix} H_1 & X \\ 0 & H_2 \end{pmatrix}$$

where H_1 is a $(k-1) \times (k-1)$ Hessenberg matrix and H_2 is an $(n-k+1) \times (n-k+1)$ Hessenberg matrix. Expanding $|A_{m+2} - \lambda I|$ by the kth row shows that it can be factorised as $|H_1 - \lambda I| \, . \, |H_2 - \lambda I|$}.

Once the search for negligible subdiagonal elements has been completed, and the matrix deflated if appropriate, the next QR iteration is performed, the origin shifts again being determined by the 2×2 submatrix currently in the lower right-hand corner.

The double QR algorithm described above is in practice an extremely efficient method of determining the eigenvalues of a real symmetric matrix. Although it is an iterative method the number of iterations needed is very small, the average number of iterations per eigenvalue being generally less than 2. The number of iterations needed can increase when A has multiple eigenvalues or a cluster of nearly equal eigenvalues, but even in these cases it is extremely rare for convergence not to occur in a reasonable number of iterations. The same algorithm can be applied to determine the eigenvalues of a symmetric tridiagonal matrix. In this case, since only orthogonal similarity transformations are used, the symmetry of the matrix is preserved and the computer storage requirement and computation time are accordingly reduced. The only disadvantage of the algorithm is that the order in which the eigenvalues are produced is somewhat arbitrary. If, for instance, the largest eigenvalues of a matrix are required, all of the eigenvalues must still be computed.

6.2.6 Eigenvector Computation

Once the eigenvalues of the upper Hessenberg matrix H have been computed by the QR algorithm, the problem of determining the eigenvectors remains. If only a few eigenvectors are required these can be found by inverse iteration with the Hessenberg matrix; for complex eigenvalues, this will entail complex arithmetic (or possibly a procedure similar to that suggested by question 5 at the end of this chapter might be adopted). If all the eigenvectors are required, it may be simpler to compute the product of all the orthogonal matrices involved in the QR iterations. This product is an orthogonal matrix which reduces the Hessenberg matrix to a nearly triangular form (pairs of eigenvalues corresponding to real 2×2 matrices along the diagonal). The eigenvectors of this nearly triangular matrix can be quickly computed by inverse iteration, and the eigenvectors of H are obtained from these via the computed product. Whichever method is adopted to find the eigenvectors of H, those of the original matrix A are computed as described in Section 6.2.3.

6.2.7 Other Methods for Non-symmetric Matrices

Attempts to adapt the Jacobi algorithm for application to non-symmetric matrices have not been particularly successful. Pre-dating the QR algorithm by a few years was the *LR algorithm* developed by H. Rutishauser in the 1950's. This is somewhat similar to the QR algorithm but depends upon factorising A as a product L_1R and defining A_1, a similar matrix, as RL_1; where, L_1 is a unit lower triangular matrix and R is again a right triangular matrix. As with the QR algorithm, origin shifts can be used to accelerate convergence, and if the iterative process starts with a Hessenberg matrix all similar matrices in the sequence are of the same form. The algorithm has the unfortunate disadvantage that not all matrices have an LR factorisation. Another method is to reduce the matrix to a tridiagonal form, but this is not a very stable process since no interchanges can be used in the second stage of the process (which is the reduction of a Hessenberg matrix to non-symmetric tridiagonal form by elementary transformations). As with the Givens–Householder method, the characteristic polynomial $f_n(x)$ can now be easily computed for any value of x by means of recurrence relations between the minors, but the Sturm sequence property is not applicable to non-symmetric tridiagonal matrices.

6.3 The Generalised Eigenvalue Problem

Frequently in engineering and scientific calculations, eigenvalue problems occur in the generalised form $A\mathbf{x} = \lambda B\mathbf{x}$ rather than in the standard form $A\mathbf{x} = \lambda\mathbf{x}$.

If either A or B is a non-singular matrix the generalised problem can be reduced to a standard form as either $(B^{-1}A)\mathbf{x} = \lambda\mathbf{x}$ or $(A^{-1}B)\mathbf{x} = \frac{1}{\lambda}\mathbf{x}$. These methods of reduction have the disadvantage that if A and B are symmetric matrices the reduced problem will generally be non-symmetric. In this symmetric case, if one of the matrices is positive definite an alternative reduced problem is available which retains the symmetry. Consider, for example, the eigenvalue problem $A\mathbf{x} = \lambda B\mathbf{x}$, where B is positive definite. Then B has a Cholesky decomposition $B = LL'$, where L is a non-singular triangular matrix. The equation is re-written as

$$A\mathbf{x} = \lambda LL'\mathbf{x}$$

whence
$$A(L')^{-1}L'\mathbf{x} = \lambda LL'\mathbf{x}$$

and so,
$$L^{-1}A(L^{-1})'(L'\mathbf{x}) = \lambda L'\mathbf{x}.$$

Letting $\mathbf{y} = L'\mathbf{x}$ and $L^{-1}A(L^{-1})' = C$, this gives the standard symmetric eigenvalue problem:–

$$C\mathbf{y} = \lambda\mathbf{y}.$$

The eigenvalues of this reduced problem are identical to those of the original problem; if $\mathbf{y}_i$ is an eigenvector of the reduced problem, the corresponding eigenvector of the original problem is $\mathbf{x}_i = (L')^{-1}\mathbf{y}_i$.

If both matrices occurring in the generalised problem are singular it is still possible that there are well-conditioned eigenvalues and eigenvectors, but none of the above methods of standardising the problem can be employed. To solve the problem in this case Moler and Stewart [5] have devised the *QZ algorithm.* Their algorithm is closely related to the *QR* algorithm and uses Householder transformations with the ultimate aim of simultaneously reducing A and B to upper triangular form. Once the triangular forms are achieved, the eigenvalues are $\lambda_i = a_{ii}/b_{ii}$, provided $b_{ii} \neq 0$. To perform the computation B is first reduced to upper triangular form either by pre-multiplying by elementary transformation matrices or by orthogonal matrices. The result is $PA\mathbf{x} = \lambda PB\mathbf{x}$, where PB is upper triangular. The matrix A (or more properly PA) is then reduced, one element at a time, to upper Hessenberg form by pre-multiplying by Householder matrices. After each step, the triangular form of B is retained by post-multiplying both matrices by suitably chosen Householder matrices. The problem is now reduced to the form $A_1\mathbf{y} = \lambda B_1\mathbf{y}$, where B_1 is upper triangular and A_1 is of upper Hessenberg form. In its simplest form, the algorithm is an iterative process which transforms

$$A_1\mathbf{y} = \lambda B_1\mathbf{y}$$

to $QA_1Z\mathbf{y} = \lambda QB_1Z\mathbf{y}$ at each iteration. The matrices Q and Z are both orthogonal matrices, in fact products of Householder matrices. Q is chosen to make QA_1 upper triangular, and Z is chosen to restore the triangular form for QB_1Z. After each iterative step, the matrix on the left is an upper Hessenberg matrix, the eigenvalues are unaltered and the eigenvectors are modified by the matrix Z. As for the QR algorithm the slow convergence of the basic algorithm is accelerated by origin shifts, again based on the 2×2 matrices on the lower right-hand corner, and a double shift version of the algorithm is used to avoid complex arithmetic.

References

1. WILKINSON, J. H. *The Algebraic Eigenvalue Problem*, O.U.P., 1965.
2. MARTIN, R. S. and WILKINSON, J. H. "Similarity Reduction of a General Matrix to Hessenberg Form". *Numer. Math.*, 12, 349–368, 1960.
3. LANCZOS, C. "An Iteration Method for the Solution of the Eigenvalue Problem of Linear Differential and Integral Operation". *J. Res. Nat. Bur. Stand.*, 45, 1950.
4. PEREYRA, V. and SCHERER, G. "Eigenvalues of Symmetric Tridiagonal Matrices". *J. Inst. Maths. Applics.*, 12, 2 October 1973, 209–222.
5. MOLER, C. B. and STEWART, G. W. "An Algorithm for Generalised Matrix Eigenvalue Problems". *SIAM J. Numer. Anal.*, 10, 2, April 1973, 241–256.

Exercises

1. Find a plane rotation matrix $R(2, 4, \theta)$ in the plane of x_2 and x_4 which eliminates the element a_{42} in the matrix

$$A = \begin{pmatrix} 2 & 3 & 2 & 1 \\ 0 & -1 & 1 & 2 \\ 0 & 4 & -1 & 3 \\ 0 & 3 & 1 & 1 \end{pmatrix}.$$

[See Section 6.1.2, RA is required to have a zero element in the (4, 2) position.]

2. The matrix $A = \begin{pmatrix} 4 & 0 & -1 \\ -2 & 2 & 1 \\ 2 & 0 & 1 \end{pmatrix}$ has exact eigenvalues $\lambda_1 = 3, \lambda_2 = \lambda_3 = 2$. Use the method of inverse iteration with $\overline{\lambda}_1 = 3.001$ to find $\mathbf{v}_1$, and with $\overline{\lambda}_2 = 1.998$ to find $\mathbf{v}_2$. Show that a linearly independent $\mathbf{v}_3$ can be obtained by using the same method with $\overline{\lambda}_3 = 2.001$.

3. Find approximately the number of individual multiplications involved in:
(a) one iteration of the simple QR algorithm without origin shift applied to a full $n \times n$ matrix A,
(b) one double shift QR iteration applied to an $n \times n$ upper Hessenberg matrix H,
(c) one double shift QR iteration applied to an $n \times n$ symmetric tridiagonal matrix.

4. Use Householder similarity transformations to reduce the 4 x 4 matrix A of Section 6.2.3 to upper Hessenberg form.

5. A real matrix A has a complex eigenvalue $\lambda = \theta + i\phi$, and associated eigenvector $\mathbf{z} = \mathbf{x} + i\mathbf{y}$. Show that $\mathbf{x}$ is a real eigenvector of the matrix $A^2 - 2\theta A$; and that $\mathbf{y} = (1/\phi)(\theta\mathbf{x} - A\mathbf{x})$. What is the corresponding eigenvalue of $A^2 - 2\theta A$? How many linearly independent eigenvectors are associated with this eigenvalue?

6. Use the method suggested by Exercise 5 to find the complex eigenvector associated with the eigenvalue $2 + 4i$ of the matrix

$$A = \begin{pmatrix} 0 & -4 & 10 \\ -2 & 2 & -10 \\ 2 & 4 & 0 \end{pmatrix}$$

7. Use the Givens method to find to one decimal place the second largest eigenvalue of the matrix

$$A = \begin{pmatrix} 3 & 1 & 0 & 0 \\ 1 & -2 & 2 & 0 \\ 0 & 2 & 1 & -1 \\ 0 & 0 & -1 & 4 \end{pmatrix}$$

8. The matrix $A = \begin{pmatrix} 19 & 2 & -2 \\ 2 & 14 & -12 \\ -2 & -12 & 14 \end{pmatrix}$ has eigenvalue $\lambda_1 = 27$ associated with $\mathbf{v}_1 = (\frac{1}{3}, \frac{2}{3}, -\frac{2}{3})'$. Find a Householder transformation, and associated matrix P, such that $P\mathbf{v}_1 = (1, 0, 0)'$. Find PAP' and hence all the eigenvalues of A. (This example demonstrates a deflation method for symmetric eigenvalue problems.)

9. T is a non-symmetric $n \times n$ tridiagonal matrix. The diagonal elements are $d_i (i = 1, \ldots, n)$, the sub-diagonal elements are ℓ_1 $(i = 1, \ldots, n-1)$, and the superdiagonal elements are u_i $(i = 1, \ldots, n-1)$. Defining $f_0(x) = 1$, $f_1(x) = d_1 - x$ and $f_k(x)$ as the determinant of the first k rows and columns of $T - xI$, show that:

$$f_k(x) = (d_k - x) f_{k-1}(x) - \ell_{k-1} u_{k-1} f_{k-2}(x), \qquad \text{for } k = 2, \ldots n.$$

T_4 is a 4 x 4 tridiagonal matrix with $d_i = 2$, $i = 1, \ldots, 4$, $\ell_i = 1$ for $i = 1, \ldots, 3$ and $u_i = 2$ for $i = 1, \ldots, 3$. By evaluating $f_4(0)$ and $f_4(2)$ show that T_4 has a real eigenvalue in the range $0 < \lambda < 2$. Use a bisection method to find this eigenvalue to 1 decimal place accuracy.

10. Find the matrix produced after applying one double shift QR iteration to the matrix

$$H = \begin{pmatrix} 2 & 2 & 4 \\ 1 & 3 & -2 \\ 0 & 1 & 0 \end{pmatrix}.$$

11. Prove that, in general, the product of an upper Hessenberg matrix, and an upper triangular matrix, in either order, is an upper Hessenberg matrix. Prove that the product of two upper Hessenberg matrices differs from upper Hessenberg form only in the elements $a_{i+2,i}$, for $i = 1, \ldots, n-2$.

12. A and B are real symmetric matrices; A is positive definite. Show how the generalised eigenvalue problem $A\mathbf{x} = \lambda B\mathbf{x}$ can be re-written as a standard problem in a manner which retains the symmetry.

CHAPTER 7

ERRORS IN EIGENVALUES AND EIGENVECTORS

7.1 Introduction

The present chapter will be concerned with errors encountered in the calculation of eigenvalues and eigenvectors. It will be brief, not because the topic is unimportant, but rather because it is so large that anything more than a nodding acquaintance with it will necessarily entail consulting a longer work, such as Dr. Wilkinson's book (reference 1).

Most of the chapter will be devoted to a consideration of the effect of perturbations and of conditioning, no more than a summary of some results being given for the effect of rounding errors. As a preliminary, we shall state and prove in outline two theorems due to S. Gerschgorin (1931).

7.2 Gerschgorin's Theorems

7.2.1 Theorem 1

Every eigenvalue of the matrix $A = (a_{ij})$ lies inside at least one of the circles in the complex plane with centre a_{ii} and radius $r_i = \sum_{j \neq i} |a_{ij}|$.

Proof

Let λ be an eigenvalue of A and let $\mathbf{x}$ be the corresponding eigenvector, so that

$$A\mathbf{x} = \lambda\mathbf{x} \tag{1}$$

Let x_i be the greatest in modulus of the components of $\mathbf{x}$. The ith equation of (1) is

$$\sum_{j=1}^{n} a_{ij} x_j = \lambda x_i.$$

Hence,

$$|(\lambda - a_{ii})x_i| \leqslant |\sum_{j \neq i} a_{ij} x_j| \leqslant \sum_{j \neq i} |a_{ij}||x_j|,$$

and therefore,

$$|\lambda - a_{ii}| \leqslant \sum_{j \neq i} \{|x_j|/|x_i|\} |a_{ij}| \leqslant \sum_{j \neq i} |a_{ij}|,$$

since

$$|x_j| \leqslant |x_i| \quad \text{for all } j \neq i.$$

7.2.2 Theorem 2

If s of the circles referred to in Theorem 1 form a connected region – isolated from the other circles – then there are exactly s eigenvalues of A in this region.

We shall not give details of this proof, but in outline it is as follows. Let D be the leading diagonal of A, and let $C = A - D$: that is

$$d_{ii} = a_{ii}, \quad d_{ij} = 0 \quad \text{for } i \neq j \text{ and}$$
$$c_{ii} = 0, \quad c_{ij} = a_{ij} \quad \text{for } i \neq j.$$

Consider the matrix $(D + \epsilon C)$, where ϵ will be treated as a variable between 0 and 1. Let Γ_i denote the Gerschgorin disc with centre a_{ii} and radius $r_i = \sum_{j \neq i} |a_{ij}|$. The coefficients of the characteristic polynomial of $(D + \epsilon C)$ are polynomials in ϵ, and it is proved in texts on algebra that therefore the zeroes of the characteristic polynomial (the eigenvalues of $D + \epsilon C$) are continuous functions of ϵ. Consequently, as ϵ increases from zero (when the eigenvalues lie at the centres of the Γ_i) the eigenvalues are confined each to its own disc for as long as the discs do not overlap. As soon as a pair of discs overlap, we can only conclude that the eigenvalues originally contained in the separate discs of the pair are now contained in the joint region covered by them. And so on as ϵ increases further and more discs coalesce. For a complete proof see, for example, reference 1.

7.3 Nearly Diagonal Matrices

7.3.1 Eigenvalues of Nearly Diagonal Matrices

If the diagonal elements of a matrix are large compared with its off-diagonal elements, Gerschgorin's theorems afford useful estimates of the eigenvalues of the matrix. For example, the eigenvalues of the matrix

$$\begin{pmatrix} 10 & 4 & 1 \\ 1 & 10 & 0.5 \\ 1.5 & -3 & 20 \end{pmatrix}$$

are $\lambda_1 = 8$, $\lambda_2 = 12$ and $\lambda_3 = 20$. The discs are

$$\Gamma_1 : \text{centre } (10, 0) \text{ radius } 5,$$
$$\Gamma_2 : \text{centre } (10, 0) \text{ radius } 1.5, \text{ and}$$
$$\Gamma_3 : \text{centre } (20, 0) \text{ radius } 4.5.$$

Γ_3 is isolated from the other discs, and therefore contains exactly one eigenvalue: λ_3. Of the other two discs, since Γ_1 includes Γ_2 we can only sat that Γ_1 contains two eigenvalues: λ_1 and λ_2. In fact, λ_1 and λ_2 both lie outside Γ_2, which is empty. Note that since the eigenvalues of a matrix are the same as that of its transpose, we can derive the radii of discs from column, instead of row, sums of the matrix. We have in this example the alternative discs

$$\pi_1 : \text{centre } (10, 0) \text{ and radius } 2.5,$$
$$\pi_2 : \text{centre } (10, 0) \text{ and radius } 7,$$
$$\pi_3 : \text{centre } (20, 0) \text{ and radius } 1.5.$$

π_3 gives a better bound for λ_3 than does Γ_3. To estimate the other two eigenvalues, we should have to use π_2, which gives worse bounds than does Γ_1. In this case, the smaller of the two circles centre (10, 0) in fact contains both λ_1 and λ_2, but we could not deduce this without knowing their values.

7.3.2 Modified Gerschgorin Discs

We can improve the bounds obtained in the last section by a simple device. To obtain an improved bound for the eigenvalue corresponding to a diagonal element *which is different from all other diagonal elements*, we multiply the row and the column containing the element by ω and by $1/\omega$ respectively, where ω will be chosen to make the Gerschgorin disc as small as possible. The matrix obtained is similar to the original matrix, and therefore has the same eigenvalues. We illustrate the technique by considering again the example used in the previous section.

We have seen that there is an eigenvalue which satisfies

$$|20 - \lambda_3| < 1.5$$

We shall attempt to improve this bound by forming the similar matrix

$$\begin{pmatrix} 10 & 4 & 1/\omega \\ 1 & 10 & 0.5/\omega \\ 1.5\omega & -3\omega & 20 \end{pmatrix}$$

The disc with centre (20, 0) is isolated from the other two discs provided that

$$20 - 1.5\omega - 3\omega > \max\{10 + 4 + 1/\omega,\ 10 + 1 + 0.5/\omega\}.$$

This is certainly satisfied if

$$0.2 \leqslant \omega \leqslant 1.2.$$

Setting $\omega = 0.2$, we find that λ_3 satisfies

$$|20 - \lambda_3| < 1.5\omega + 3\omega = 1.35.$$

Using column instead of rows, we do even better. We now require $1/\omega$ to lie in the range $0.6 < 1/\omega < 3.4$, and choosing the lower value:

$$|20 - \lambda_3| < 1/\omega + 0.5/\omega = 0.9.$$

7.3.3 Eigenvalues of a Perturbed Diagonal Matrix

We shall deal with the eigenvalues of a perturbed diagonal matrix. It will become clear subsequently that an analysis of this very special case is of wider relevance.

We suppose that there is a diagonal matrix D, with diagonal elements (and hence also eigenvalues) $d_1, d_2, \ldots, d_n$; and which is perturbed by a small matrix which we denote by ϵP. Here, ϵ is the modulus of the largest element of the perturbation and hence $|p_{ij}| \leqslant 1$ for all i, j. Denote the perturbed matrix by C so that

$$C = D + \epsilon P.$$

We are concerned to estimate by how much the eigenvalues of C differ from those of D.

If λ is any eigenvalue of C, then for some i:

$$|\lambda - c_{ii}| < \epsilon \sum_{j \neq 1} |p_{ij}| \leqslant (n-1)\epsilon,$$

and hence

$$|\lambda - d_i| = |\lambda - (c_{ii} - p_{ii})| \leqslant (n-1)\epsilon + |p_{ii}| \leqslant n\epsilon.$$

Every eigenvalue of C lies in one of the circles

$$|\lambda - d_i| \leqslant n\epsilon \qquad \text{for } i = 1, 2, \ldots, n; \tag{3}$$

and hence – from theorem 2 – if S of the circles defined by (3) form a connected region, then that region includes S of the eigenvalues of C.

If a particular diagonal element of C is different from all its other diagonal elements, then for sufficiently small ϵ the disc with that element as centre is isolated from all other discs. For simplicity, suppose that the element referred to is c_{11}. Then provided

$$\epsilon < \delta/2n, \qquad \text{where } \delta = \min_{i=2,3\ldots n} |c_{11} - c_{ii}|, \tag{4}$$

we have an eigenvalue of C – say λ_1 – which satisfies

$$|\lambda_1 - d_1| < n\epsilon. \tag{5}$$

Note that λ_1 is necessarily a simple eigenvalue of C, as is d_1 of D. Note also that we may take δ in (4) as $\min_{j=2,3\ldots n} |d_1 - d_j|$ with negligible error.

If the diagonal elements of C are all distinct, then for sufficiently small ϵ, the eigenvalues of D and of C can be paired off so that

$$|\lambda_i - d_i| \leqslant n\epsilon \qquad \text{for } i = 1, 2, \ldots, n. \tag{6}$$

These bounds are probably satisfactory provided that the elements of D are not too different in magnitude, but notice that otherwise the *relative* changes in eigenvalues due to perturbations are not necessarily small. Consider as an example

$$D = \begin{pmatrix} 1 & 0 \\ 0 & 0.01 \end{pmatrix}, \qquad P = \begin{pmatrix} -0.1 & 1 \\ 1 & -0.1 \end{pmatrix}, \qquad \epsilon = 0.1.$$

The eigenvalues of

$$C = \begin{pmatrix} 0.99 & 0.1 \\ 0.1 & 0 \end{pmatrix}$$

are $\lambda_1 = 1$ and $\lambda_2 = -0.01$. Certainly, inequality (6) is satisfied for both eigenvalues, and the absolute changes in eigenvalues are small. On the other hand, the eigenvalue $d_2 = 0.01$ is changed by 200%. We would stress that the point made here can be of practical importance. The matrix D, with two positive eigenvalues, may correspond to a stable physical system; in which case matrix C, with one negative eigenvalue, will correspond to an unstable system.

7.3.4 Improved Bounds for Eigenvalues

We can employ the device used in Section 7.3.2 to decrease the bound given in (6). Again, let c_{11} be different from all other c_{ii}, and multiplying the first row of C by ω and its first column by $1/\omega$: say

$$Q = \begin{pmatrix} c_{11} & \epsilon\omega p_{12} & \cdots\cdots & \epsilon\omega p_{1n} \\ \epsilon p_{21}/\omega & c_{22} & \cdots\cdots & \epsilon p_{2n} \\ \cdot & & & \\ \cdot & & & \\ \epsilon p_{n1}/\omega & \epsilon p_{n2} & \cdots\cdots & c_{nn} \end{pmatrix}$$

The first Gerschgorin disc of Q is

$$\Gamma_1 : |c_{11} - \lambda| \leqslant \epsilon\omega \sum_{j=2}^{n} |p_{1j}| < (n-1)\epsilon\omega,$$

whilst the remaining discs are

$$\Gamma_i : |c_{ii} - \lambda| \leqslant \epsilon |p_{i1}|/\omega + \sum_{j \neq 1,i} \epsilon |p_{ij}| \leqslant \epsilon/\omega + (n-2)\omega.$$

Discs Γ_1 and Γ_i do not overlap provided

$$|c_{11} - c_{ii}| > (n-1)\epsilon\omega + \epsilon/\omega + (n-2)\epsilon. \tag{7}$$

If we set $\eta_i = |c_{11} - c_{ii}|/\epsilon$, then inequality (7) is satisfied if

$$(n-1)\omega^2 - (\eta_i - n + 2)\omega + 1 < 0. \tag{8}$$

It is easy to show that if ϵ satisfies inequality (4), then (8) may be satisfied by real values of ω, and the lower limit of permissible values of ω is given by

$$2(n-1)\omega \geqslant (\eta_i - n + 2) - \sqrt{\{(\eta_i - n + 2)^2 - 4(n-1)\}}. \tag{9}$$

The right-hand side of (9) is less than or equal to $4(n-1)/(\eta_i - n + 2)$, so that if we choose

$$\omega = \max_{i=2,3,\ldots n} \{2(\eta_i - n + 2)^{-1}\} = 2\epsilon(\delta - n\epsilon + 2\epsilon)^{-1}, \tag{10}$$

where δ is given by (4), then inequality (9) is satisfied for all $i \neq 1$ and hence Γ_1 is isolated from all other Γ_i. It follows then that there is an eigenvalue of C – say λ_1 – which satisfies the inequality

$$|c_{11} - \lambda_1| \leqslant \frac{2(n-1)\epsilon^2}{(\delta - n\epsilon + 2\epsilon).} \tag{11}$$

Notice that we have already demanded – inequality (4) – that ϵ should be less than $\delta/2n$, and consequently the right-hand side of (11) is at most $2(n-1)\epsilon/(n+2)$, so that approximately

$$|c_{11} - \lambda_1| \leqslant 2\epsilon. \tag{12}$$

Provided that the eigenvalues are well spaced, we may consider the case when ϵ is small compared with δ/n, and we have approximately

$$|c_{11} - \lambda_1| \leqslant 2(n-1)\epsilon^2 / \min_{i=2,3\ldots n} |d_1 - d_i|. \tag{13}$$

(The difference between $|c_{11} - c_{ii}|$ and $|d_1 - d_i|$ introduces a term of higher order in ϵ which we suppose to be negligible.) Notice that unless we have special information about the perturbation of the diagonal elements, the difference between λ_1 and d_1 is still of order ϵ – rather than of order ϵ^2. We have

$$|\lambda_1 - d_1| = |\lambda_1 - (c_{11} - \epsilon p_{11})| \leqslant |\lambda_1 - c_{11}| + \epsilon |p_{11}|, \text{ and hence}$$

$$|\lambda_1 - d_1| \leqslant \epsilon + 2(n-1)\epsilon^2 / \min_{i=2\ldots n} |d_1 - d_i|. \tag{14}$$

On the same assumption – of ϵ small compared with δ/n – we then have

$$|\lambda_1 - d_1| \leqslant \epsilon. \tag{15}$$

On the assumption only that inequality (4) is satisfied we obtain from (12)

$$|\lambda_1 - d_1| < 3\epsilon. \tag{16}$$

7.3.5 Eigenvectors of a Perturbed Diagonal Matrix

We consider now the effect of a small perturbation on an eigenvector of a diagonal matrix. We suppose that the eigenvector corresponds to a simple eigenvalue of D, and for convenience take this eigenvalue to be d_1. Let ϵ satisfy inequality (4) so that there is a simple eigenvalue, λ_1, of C which satisfies inequality (14).

Since d_1 is a simple eigenvalue of D, its corresponding right eigenvector is

$$\mathbf{e}_1 = (1, 0, \ldots, 0)'.$$

We are concerned to discover by how much the eigenvector of C which corresponds to λ_1 (say $\mathbf{u}$) differs from $\mathbf{e}_1$. If we normalise $\mathbf{u}$ by making its first element equal to unity, we hope to show that the other elements are small. We have

$$C\mathbf{u} = (D + \epsilon P)\mathbf{u} = \lambda_1 \mathbf{u}. \tag{17}$$

It is convenient to partition the matrices in equation (17) thus:

$$D = \left(\begin{array}{c|c} d_1 & \mathbf{0}' \\ \hline \mathbf{0} & D_1 \end{array}\right), \qquad P = \left(\begin{array}{c|c} p_{11} & \boldsymbol{\alpha}' \\ \hline \boldsymbol{\beta} & P_1 \end{array}\right), \qquad \mathbf{u} = \left(\begin{array}{c} 1 \\ \hline \mathbf{u}_1 \end{array}\right).$$

The equations (17) are of rank $(n-1)$, so that we may drop the first equation and solve the remaining $(n-1)$ for $\mathbf{u}_1$. We have

$$D_1\mathbf{u}_1 + \epsilon(\boldsymbol{\beta} + P_1\mathbf{u}_1) = \lambda_1\mathbf{u}_1,$$

which we can rewrite as

$$(\lambda_1 I - D_1 - \epsilon P_1)\mathbf{u}_1 = \epsilon\boldsymbol{\beta}. \tag{18}$$

We can consider equation (18) as the equation

$$(\lambda_1 I - D_1)\mathbf{u}_1 = \mathbf{0},$$

when the matrix of coefficients is perturbed by the matrix ϵP_1, and the right-hand side by the vector $\epsilon\boldsymbol{\beta}$. We require that $(\lambda_1 I - D_1)$ should not be ill-conditioned for inversion. This consideration leads to the requirement that d_1 should not be close to any other element of D. We shall use equation (18) to construct a bound for $\mathbf{u}_1$, and show that $\mathbf{u}_1$ is small if d_1 is well separated from other elements (eigenvalues) of D.

Since d_1 is a simple eigenvalue of D and satisfies inequality (4), λ_1 is not equal to d_i for any $i = 2, 3 \ldots n$. It follows that $(\lambda_1 I - D_1)$ is not singular. Define Q_1 as the diagonal matrix

$$Q_1 = \text{diag}\,([\lambda_1 - d_i]^{-1}) = (\lambda_1 I - D_1)^{-1}.$$

Pre-multiplying equation (18) by Q_1 gives

$$(I - \epsilon Q_1 P_1)\mathbf{u}_1 = \epsilon Q_1 \boldsymbol{\beta}.$$

For any matrix M with norm less than unity, $(I - M)$ is not singular, and $\|(I - M)^{-1}\|$ is not greater than $(1 - \|M\|)^{-1}$ (see Appendix). But

$$\|Q_1\| = \max_{i=2,\ldots,n} |\lambda_1 - d_i|^{-1} \approx \delta^{-1} \tag{19}$$

where δ is given by (4), and hence

$$\|\epsilon Q_1 P_1\| \leqslant \epsilon \|Q_1\|\,\|P_1\| \leqslant \epsilon\delta^{-1}(n-1) < 1$$

since ϵ satisfies inequality (4). It follows that we may write

$$\mathbf{u}_1 = (I - \epsilon Q_1 P_1)^{-1} \epsilon Q_1 \boldsymbol{\beta}$$

and hence

$$\|\mathbf{u}_1\| \leqslant (1 - \|\epsilon Q_1 P_1\|)^{-1} \|\epsilon Q_1 \boldsymbol{\beta}\|$$

$$\leqslant \frac{\epsilon \|Q_1\|\,\|\boldsymbol{\beta}\|}{1 - \epsilon \|Q_1\|\,\|P_1\|} \tag{20}$$

With negligible difference, we can write $\|Q_1\| = \delta^{-1}$, and since $\|P_1\| \leqslant (n-1)$

$$\|\mathbf{u}_1\| \leqslant \frac{\epsilon \|\boldsymbol{\beta}\|}{\delta - (n-1)\epsilon}.$$

Since each element of $\boldsymbol{\beta}$ is less than unity in modulus, if we let g denote the norm of a column of ones, we have for each norm

$$\|\mathbf{u}_1\| \leqslant \frac{\epsilon}{\delta - (n-1)\epsilon} \cdot g. \tag{21}$$

which (since $x_1 = 1$) is satisfactory provided $\epsilon / [\delta - (n-1)\epsilon]$ is small. Note that when considering eigenvectors, the bound obtained is probably not satisfactory if ϵ is close to $\delta/2n$. In this case, the right-hand side of (21) approaches $g/(n+1)$, which for ℓ_1, ℓ_2 and ℓ_∞ norms is equal to $(n-1)/(n+1)$. We require that ϵ should be small compared with δ/n, and may therefore fear large errors if d_1 is close to some other eigenvalue of D so that δ is large.

A simple example will illustrate this. The eigenvalues of the matrix

$$C = \begin{pmatrix} 1 & 0.01 & 0.01 \\ 0 & 1.01 & 0.01 \\ 0.01 & -0.01 & 0.5 \end{pmatrix}$$

are 1.0, 1.01 and 0.5. The corresponding eigenvectors are

$$\begin{pmatrix} 1 \\ 1/\sqrt{51} \\ 1/\sqrt{51} \end{pmatrix}, \quad \begin{pmatrix} 1 \\ 1 \\ 0 \end{pmatrix} \quad \text{and} \quad \begin{pmatrix} -1/\sqrt{51} \\ -1/\sqrt{51} \\ 1 \end{pmatrix}.$$

The first and third of these are close to the unperturbed eigenvectors $(1, 0, 0)'$ and $(0, 0, 1)'$, but the second eigenvector is seriously different from $(0, 1, 0)'$.

We shall not consider the case of repeated or multiple eigenvalues in any detail, but we may note that if the diagonal matrix D has two equal elements – say d_1 and d_2 – then any vector of the form

$$\mathbf{x}(\alpha) = \begin{pmatrix} 1 \\ 0 \\ 0 \\ \vdots \\ 0 \end{pmatrix} + \alpha \begin{pmatrix} 0 \\ 1 \\ 0 \\ \vdots \\ 0 \end{pmatrix} \tag{22}$$

is an eigenvector. The perturbed matrix will, in general, not have corresponding equal eivengalues, nor will its corresponding eigenvectors necessarily be close to either $(1\,0\,0\,\ldots\,0)'$ or to $(0\,1\,0\,0\,\ldots\,0)'$; but we may expect the perturbed eigenvectors to be close to $\mathbf{x}(\alpha)$ for some pair of values of α. For example, we may regard the matrix given above as a diagonal matrix with two equal elements plus a small perturbation, thus:

$$C = \begin{pmatrix} 1 & 0 & 0 \\ 0 & 1 & 0 \\ 0 & 0 & 0.5 \end{pmatrix} + 0.01 \begin{pmatrix} 0 & 1 & 1 \\ 0 & 1 & 1 \\ 1 & -1 & 0 \end{pmatrix}.$$

Ignoring terms of order ϵ, the eigenvectors of C corresponding to the two eigenvalues close to $\lambda = 1$ are

$$\mathbf{x}_1 = \begin{pmatrix} 1 \\ 0 \\ 0 \end{pmatrix} \quad \text{and} \quad \mathbf{x}_2 = \begin{pmatrix} 1 \\ 1 \\ 0 \end{pmatrix}.$$

The eigenvectors of the diagonal matrix are given equally well by

$$\mathbf{x} = \begin{pmatrix} 1 \\ 0 \\ 0 \end{pmatrix} + \beta \begin{pmatrix} 1 \\ 1 \\ 0 \end{pmatrix} \quad \text{for some } \beta$$

as by equation (16). For further details see reference 1, p. 75.

7.4 General Perturbed Matrices

We consider now the effect of perturbations on the eigenvalues and eigenvectors of a matrix which is not necessarily diagonal. We assume that the matrix under consideration can be reduced to a diagonal matrix by a similarity transformation. For a consideration of other cases see reference 1.

7.4.1 Eigenvalues of a Perturbed Matrix

We first consider an example. The matrix

$$A = \begin{pmatrix} 9 & 4.5 & 3 \\ -56 & -28 & -18 \\ 60 & 30 & 19 \end{pmatrix} \tag{23}$$

has eigenvalues 1, 0, −1; but the matrix

$$A + B = \begin{pmatrix} 9 & 4.5 & 3 \\ -56 & -28 & -18 \\ 60 & 30 & 19 \end{pmatrix} + 0.05 \begin{pmatrix} 0 & 0 & 0 \\ 0 & 0 & 0 \\ 0 & 0 & -1 \end{pmatrix}$$

has eigenvalues 0.2, 0, −0.25. One of the eigenvalues is unchanged, but the other two have been changed by 80% and 75% respectively as the result of a perturbation which is only about 0.01% of the magnitude of A itself. A diagonal matrix with the same eigenvalues as A would be only very slightly affected by such a relative perturbation, and the ill-conditioning of the eigenvalues of A is due to something other than closeness of eigenvalues. We can see the reason for this ill-conditioning if we consider the reduction of A to diagonal form. Taking X as the square matrix with right-eigenvectors of A as columns:

$$X = \begin{pmatrix} 3 & 1 & 0 \\ -12 & -2 & 2 \\ 10 & 0 & -3 \end{pmatrix},$$

we have

$$D = X^{-1}AX = \begin{pmatrix} 1 & & \\ & 0 & \\ & & -1 \end{pmatrix}.$$

The same transformation in the case of δA gives

$$\delta D = X^{-1}BX = \begin{pmatrix} -0.5 & 0 & 0.15 \\ 1.5 & 0 & -0.45 \\ -1.5 & 0 & 0.45 \end{pmatrix}$$

The elements of δD are as much as thirty times as large as the largest of those of B, and we may relate the comparatively large change in eigenvalues to this increase in magnitude. We shall examine this point more formally in the next section.

Note that the matrix A itself is in very special relationship with X, and the similarity transformation does not magnify its elements. In fact, in this case $\|A\|_\infty = 109$ whilst $\|D\|_\infty = 1$. This large decrease in magnitude, together with the magnification of the perturbation accounts for the large change consequent upon such a small *relative* change.

7.4.2 A Condition Number for the Eigenvalue Problem

We consider first the overall change in the eigenvalues of A due to a perturbation B. Let λ be an eigenvalue of $(A + B)$, so that $(A + B - \lambda I)$ is singular, and hence also the matrix.

$$F = X^{-1}(A + B - \lambda I)X = D - \lambda I + X^{-1}BX.$$

If we assume for the time being that λ is different from the eigenvalues of A then the matrix

$$Q = D - \lambda I$$

is non-singular, and we can write F in the form

$$F = Q(I + Q^{-1}X^{-1}BX).$$

Since F is singular and Q is not, it follows that $(I + Q^{-1}X^{-1}BX)$ is singular and that therefore

$$\|Q^{-1}X^{-1}BX\| > 1, \tag{24}$$

(since otherwise $(I + Q^{-1}X^{-1}BX)$ would be non-singular – see Appendix). The matrix Q^{-1} is diagonal with elements $(d_i - \lambda)^{-1}$, and hence for each of the ℓ_1, ℓ_∞ and ℓ_2 norms

$$\|Q^{-1}\| = \max_i |(d_i - \lambda)^{-1}| = (\min_i |d_i - \lambda|)^{-1}.$$

From (24),

$$1 < \|Q^{-1}\|\ \|X^{-1}\|\ \|B\|\ \|X\| = (\min_i |d_i - \lambda|)^{-1}\ \|X^{-1}\|\ \|X\|\ \|B\|,$$

and finally

$$\min_i |d_i - \lambda| < \|X^{-1}\|\ \|X\|\ \|B\|.$$

In deriving this inequality we assumed that λ was different from each d_i, but clearly it also holds if λ is equal to d_i for some i. We deduce, then, that for some i

$$|\lambda - d_i| < \|X^{-1}\|\ \|X\|\ \|B\|. \tag{25}$$

The matrix X in (25) is not unique (its columns may each be multiplied by a constant and (25) holds for any X such that $X^{-1}AX$ is diagonal. It follows, therefore, that for each eigenvalue, λ, of $(A + B)$ there is at least one eigenvalue, λ_i, of A such that

$$|\lambda - d_i| < K\|B\|, \quad \text{where} \tag{26}$$
$$K = \min(\|X^{-1}\| \, \|X\|),$$

and the minimum is taken over all X such that $X^{-1}AX$ is diagonal. The number K affords a condition number for the evaluation of the eigenvalues of A. If the ℓ_2 norm is used, K is referred to as the spectral condition number. We may note that A is ill-conditioned for this problem if the matrix X (however scaled) is ill-conditioned for inversion. Note also that the bound given by (26) makes no assumption about the magnitude of A; it is a bound expressed in terms of the absolute perturbation of A rather than of a relative perturbation. We have seen that the bound could be a good deal worse if expressed in terms of a relative perturbation. No assumption has been made that B is small, nor that λ corresponds to a simple eigenvalue.

7.4.3 Perturbations of a Simple Eigenvalue

Not all the eigenvalues of a matrix are affected in the same way by a perturbation. In Section 7.4.1, for example, we saw two eigenvalues changed drastically whilst the third was unaltered. In fact, the conditioning of the separate eigenvalues may be quite different, and we construct now a bound for the change in a specific (simple) eigenvalue.

We choose the matrix X to have, as columns, normalised eigenvectors – say $\mathbf{x}_i$ so that $\|\mathbf{x}_i\|_2 = 1$ for each i. If $\mathbf{y}_i'$ denotes the corresponding normalised row eigenvectors ($\|\mathbf{y}_i\|_2 = 1$), then the rows of X^{-1} are $k_i\mathbf{y}_i'$, where $k_i = (\mathbf{y}_i'\mathbf{x}_i)^{-1}$. As before, consider the matrix

$$X^{-1}(A + B)X = D + X^{-1}BX = D + Q, \text{ say.}$$

The elements of Q are given by

$$q_{ij} = k_i\mathbf{y}_i'B\mathbf{x}_j, \quad \text{so that}$$
$$|q_{ij}| \leq |k_i| \, \|\mathbf{y}_i'\|_2 \, \|B\|_2 \, \|\mathbf{x}_j\|_2 = |k_i| \, \|B\|_2 . \tag{27}$$

The eigenvalues of $(D + Q)$ – which are the same as those of the perturbed matrix $(A + B)$ – lie within the Gerschgorin discs centre $(d_i + q_{ii})$ and radius $\sum_{j \neq i} |q_{ij}|$. If we consider an isolated eigenvalue of A – say d_i – then for $\|B\|_2$ sufficiently small the disc with centre $(d_i + q_{ii})$ is isolated and hence contains exactly one eigenvalue of $(A + B)$ – say λ – which satisfies

$$|\lambda - (d_i + q_{ii})| \leq \sum_{j \neq i} |q_{ij}|,$$

and hence, using (22),

$$|\lambda - d_i| \leqslant \sum_{j=1}^{n} |q_{ij}| < n|k_i|\,\|B\|_2. \tag{28}$$

(Note that $k_i = (\mathbf{y}_i'\mathbf{x}_i)^{-1}$ where $\mathbf{y}_i'$ and $\mathbf{x}_i$ are the normalised left and right eigenvectors corresponding to the eigenvalue d_i.) Inequality (28) is similar to the bound given in (26), but allows for a closer bound for eigenvalues which are not ill-conditioned. Clearly, the numbers $|k_i|$ afford condition numbers for the individual eigenvalues.

We can derive a bound corresponding to that given in (14) if we write

$$X^{-1}(A + B)X = D + Q = D + \epsilon P, \quad \text{where} \tag{29}$$

$$\epsilon = \max_{ij} |q_{ij}|,$$

so that $|p_{ij}| \leqslant 1$ for all i, j. We can now use the analysis of Section 7.3.3. In particular, from (14)

$$|\lambda_i - d_i| \leqslant \epsilon + 2(n-1)\epsilon^2/\min_{j \neq i} |d_i - d_j|. \tag{30}$$

Assuming that no other eigenvalue is too close to d_i, and that ϵ is small, we have approximately

$$|\lambda_i - d_i| \leqslant \epsilon,$$

and using (27) and (29) we deduce

$$|\lambda_i - d_i| \leqslant \max_i (|k_i|\,\|B\|_2) = \|B_2\| \max_i |k_i|. \tag{31}$$

It is possible, of course, that for some eigenvalues the bound given by (28) is better than that given by (30).

Note that k_i is large – and hence d_i possibly sensitive to perturbations – if $\mathbf{y}_i'\mathbf{x}_i$ is small, which implies that $\mathbf{y}_i$ and $\mathbf{x}_i$ (both being of unit length) are nearly orthogonal.

7.4.4 Perturbation of Eigenvectors

Using the analysis of Section 7.3.5 with equation (29) we can write down a bound for the error in an eigenvector due to the perturbation. We consider for simplicity of notation the eigenvector, $\mathbf{u}$, corresponding to the eigenvalue, λ_1, of $(D + \epsilon P)$. We suppose that the eigenvalue d_1 of D (and of A) is simple and that ϵ (and B) is sufficiently small that λ_1 also is necessarily simple. Let $\mathbf{u}$ be normalised so that its first element is unity, and let $\mathbf{u}_1$ be the $(n-1)$ dimensional vector obtained by deleting the first element of $\mathbf{u}$. Then, from (21),

$$\|\mathbf{u}_1\| \leqslant g\epsilon/[\min_{i \neq 1} |d_1 - d_i| - (n-1)\epsilon], \tag{32}$$

where g has the same meaning as in (21). From equation (29), the eigenvector of $(A+B)$ which corresponds to its eigenvalue λ_1 is

$$\mathbf{v} = X\mathbf{u}.$$

In terms of the columns of X (which are eigenvectors of A itself) remembering that $u_1 = 1$, we can write $\mathbf{v}$ in the form

$$\mathbf{v} = \mathbf{x}_1 + u_2\mathbf{x}_2 + \ldots + u_n\mathbf{x}_n.$$

Using the ℓ_2 norm, a bound for the change in $\mathbf{x}_1$ is given by

$$\|\mathbf{v}-\mathbf{x}_1\|_2 \leqslant \|\sum_{i=2}^{n} u_i\mathbf{x}_i\|_2 \leqslant \sum_{i=2}^{n} |u_i|\,\|\mathbf{x}_i\|_2 \leqslant \sum_{i=2}^{n} |u_i|, \tag{33}$$

since the columns of X were normalised so that $\|\mathbf{x}_i\|$ was equal to unity for each i. The right-hand side of (33) is equal to $\|\mathbf{u}_1\|_1$. Using the ℓ_1 norm in (32), g is the ℓ_1 norm of a column of $(n-1)$ ones and hence equal to $(n-1)$. Inequality (33) becomes, therefore,

$$\|\mathbf{v}-\mathbf{x}_1\|_2 \leqslant \frac{(n-1)\epsilon}{\min\limits_{i\neq 1} |d_1-d_i| - (n-1)\epsilon} \tag{34}$$

where – from (27) and (29)

$$\epsilon \leqslant \|B\|_2 \max_i |k_i|. \tag{35}$$

For (34) to be a satisfactory bound, it is necessary that ϵ should be small compared with $(n-1)^{-1} \min\limits_{i\neq 1} |d_1-d_i|$, and we, therefore, require

$$\|B\|_2 \ll \frac{\min\limits_{i\neq 1} |d_1-d_i|}{(n-1)\max\limits_i |k_i|} \tag{36}$$

If this inequality is satisfied then, from (34),

$$\|\mathbf{v}-\mathbf{x}_1\|_2 \ll 1,$$

which is clearly a satisfactory result since $\|\mathbf{x}_1\|_2$ is equal to unity. On the other hand, the eigenvector corresponding to any simple eigenvalue d_j, may be ill-conditioned if d_j is close to any other eigenvalue or if *any* of the left-hand eigenvalue pairs are nearly parallel.

7.4.5 Transformations and Properties of Condition Numbers

The condition number K and the individual condition numbers k_i are all greater than or equal to unity in modulus. We have for any X and any norm

$$\|X^{-1}\|\,\|X\| \geqslant \|X^{-1}X\| = \|I\| \geqslant 1,$$

and hence $K \geqslant 1$. Also for any vectors $\mathbf{x}$ and $\mathbf{y}$,

$$|\mathbf{y}'\mathbf{x}| \leqslant \|y\|_2 \; \|x\|_2,$$

so that if $\|\mathbf{y}_i\|_2 = \|x_i\|_2 = 1$, then

$$|\mathbf{y}_i'\mathbf{x}_i| \leqslant 1, \qquad \text{and hence } |k_i| \geqslant 1.$$

We consider now the effect of a similarity transformation on the condition number of a matrix. Let A be diagonalised by the similarity transformation

$$D = X^{-1}AX.$$

Consider the matrix obtained from A by any arbitrary transformation:

$$A_1 = HAH^{-1}, \qquad A = H^{-1}A_1 H.$$

Then

$$(HX)^{-1}A_1(HX) = D.$$

The bounds for the eigenvalues of $(A_1 + B)$ are, from (25),

$$|\lambda - d_i| < \|(HX)^{-1}\| \; \|XH\| \; \|B_1\|. \tag{37}$$

For an arbitrary matrix H, the best bound we can give is

$$|\lambda - d_i| < \|H^{-1}\| \; \|X^{-1}\| \; \|X\| \; \|H\| \; \|B_1\|,$$

which gives the condition number for A_1 as

$$K_1 = \min \|H\| \; \|H^{-1}\| \; \|X\| \; \|X^{-1}\| = \|H\| \; \|H^{-1}\| \min \|X\| \; \|X^{-1}\|$$
$$= \|H\| \; \|H^{-1}\| \, K. \tag{38}$$

A similarity transformation may, of course, yield a matrix with better conditioning if the transformation is specially related to the transformed matrix. Nevertheless in general we can only rely on (25), and since $\|H\| \; \|H^{-1}\| \geqslant 1$ the condition number may be increased.

In one case, we can be sure that the spectral condition number, at least, does not increase; that is, when the transformation is orthogonal. This means that $H'H = I$, so that all the eigenvalues of $H'H$ are equal to unity and therefore

$$\|H\|_2 = 1. \tag{39}$$

Also, since H^{-1} is equal to H', $\|H^{-1}\|_2$ is equal to unity. Consequently, $\|H\|_2 \, \|H^{-1}\|_2$ equals unity, and therefore K_1 has the same value as K.

A similar result is obtained if we consider the individual condition numbers $|k_i|$. If $\mathbf{x}_i$ and $\mathbf{y}_i'$ are normalised eigenvectors of A corresponding to the eigenvalue d_i, then $H\mathbf{x}$ and $\mathbf{y}'H^{-1}$ are the corresponding eigenvectors of A_1. Then, if H is an orthogonal matrix:

$$\|H\mathbf{x}\|_2 = \sqrt{\mathbf{x}'H'H\mathbf{x}} = \sqrt{\mathbf{x}'\mathbf{x}} = \|\mathbf{x}\|_2 = 1,$$

and similarly $\|y'H^{-1}\|_2$ equals unity. The condition number for d_i as eigenvalue of A_1 is, therefore,

$$|k_i^{(1)}| = \mathbf{y}'H^{-1}H\mathbf{x} = \mathbf{y}'\mathbf{x} = |k_i|.$$

Again we find that an orthogonal transformation can be relied upon not to make worse the conditioning of the eigenvalue problem.

7.5 Real Symmetric Matrices

7.5.1 Condition Numbers for Real Symmetric Matrices

We show first that a real symmetric matrix is always well-conditioned with respect to its eigenvalues.

If A is a real symmetric matrix, then it can be diagonalised by an orthogonal transformation ($X'X = I$). We have seen (39) that therefore $\|X\|_2$ and $\|X^{-1}\|$ are both equal to unity, so that the spectral condition number is given by $K = 1$, its minimum possible value. Moreover, if $\mathbf{x}_i$ is a column of the orthogonal matrix X, then since $X'X = I$, we have for each i:

$$\|\mathbf{x}_i\|_2 = \sqrt{\mathbf{x}_i'\mathbf{x}_i} = 1. \tag{40}$$

The rows of the inverse matrix $X^{-1} = X'$ are also $\mathbf{x}_i$, and therefore they also are normalised. Hence the condition numbers $|k_i|$ are given by

$$|k_i| = 1/|\mathbf{y}_i'\,\mathbf{x}_i| = 1/(\mathbf{x}_i'\,\mathbf{x}_i) = 1,$$

from (40). The individual condition numbers of a real symmetric matrix take their minimum possible values, and again we see that such a matrix is perfectly conditioned.

7.5.2 Bounds for Eigenvalues of a Real Symmetric Matrix

From the previous section and inequality (25) we deduce that every eigenvalue of $(A + B)$ lies inside one of the discs

$$|d_i - \lambda| \leqslant \|B\|_2. \tag{41}$$

However, generally if a matrix is symmetric, it is known *a priori* to be symmetric and is dealt with as such, only the diagonal and one triangle being computed and stored, and consequently any perturbation will also be symmetric. In this case, we can state two stronger results than are given by (41).

If A and $A + B$ are both symmetric and have respectively eigenvalues $d_1 < d_2 < \ldots < d_n$ and $\lambda_1 < \lambda_2 < \ldots < \lambda_n$, then

$$|d_i - \lambda_i| \leqslant \|B\|_2 \qquad \text{for all } i. \tag{42}$$

Equation (42) gives the same bound as (41), but at the same time pairs off the two sets of eigenvalues without requiring that B should be small, nor that the d_i should be simple eigenvalues.

With the same assumptions, the following result is due to Hoffman and Weilandt (see, for example, Reference 1, p. 104).

$$\sum_{i=1}^{n} (d_i - \lambda_i)^2 \leqslant \|B\|_E^2. \tag{43}$$

If we suppose that all elements of B are less than ϵ in absolute magnitude, then inequality (42) gives

$$\sum_{i=1}^{n} (d_i - \lambda_i)^2 \leqslant n\ \|B\|_2^2 < n^3 \epsilon^2;$$

whilst inequality (43) gives the left-hand side as not greater than $n^2 \epsilon^2$, and provides a better bound. Note, however, that (43) does not necessarily guarantee a better bound than (42) for every eigenvalue. From (43) we can only deduce for individual eigenvalues that

$$|d_i - \lambda_i| < \|B\|_E,$$

and since $\|B\|_E$ is never less than $\|B\|_2$, this does not improve upon (42).

7.6 Rounding Errors in Eigenvalue Problems

The effects of rounding errors on the calculation of eigenvalues and eigenvectors are by no means as easy to analyse as in the case of the solution of sets of equations, and we can do no more than touch upon the problem here. Explicit bounds are given for the Givens-Householder method in the next section, but bounds are not readily available in other cases. It is, however, possible to construct bounds *a posteriori*, and this construction is described in 7.6.2.

7.6.1 Error Bounds for the Givens-Householder Method

Ortega (Ref. 2) gives bounds for the effect of rounding errors in the floating-point Givens-Householder reduction of a matrix to tridiagonal form. If the eigenvalues of a real symmetric matrix A are $d_1, d_2, \ldots, d_n$, and the eigenvalues of the computed tridiagonal matrix are $\lambda_1, \lambda_2, \ldots, \lambda_n$, then

$$\frac{\max |d_i - \lambda_i|}{\max |d_i|} \leqslant \frac{f(n)2^{-t}}{1 - f(n)2^{-t}}, \tag{44}$$

where $f(n)$ is $6n^{5/2}$ plus terms of lower power in $\sqrt{n}$, and t is the number of binary digits in the mantissa of the floating-point arithmetic.

If, for example, $t = 30$ and $n = 100$, then the right-hand side of (44) is approximately 10^{-3}. It is, of course, still necessary to find the eigenvalues of the tri-diagonal matrix, but it can be shown that the error in evaluating these eigenvalues can be made negligibly small compared with the bound given in (44), so that this bound effectively holds for the errors in the computed eigenvalues. Experience suggests that (44) is much too pessimistic, and it would probably be safe to take $f(n)$ as $\sqrt{n}$.

Assuming that $f(n)2^{-t}$ is small, the errors in the computed eigenvectors can be made small provided that the eigenvalues are well spaced.

7.6.2 *A Posteriori* Error Bounds

If computed values have been obtained for *all* the eigenvalues and eigenvectors of a matrix, we can use a technique described by Wilkinson (reference 1, p. 80) to obtain bounds for the computed values, and possibly at the same time improve them. The idea is essentially the same as used in the discussion of the effects of perturbations in the original matrix, except that the errors are now due to errors in computation and the roles of the exact and the perturbed matrix are reversed.

Let γ_i and $\mathbf{v}_i$ $(i = 1, 2, \ldots, n)$ be approximate eigenvalues and eigenvectors of a matrix A. Let Γ be the diagonal matrix with elements γ_i, and V the square matrix with columns $\mathbf{v}_i$. If the γ_i and $\mathbf{v}_i$ were exact eigenvalues and eigenvectors we should have $AV = V\Gamma$, so that if we form the matrix

$$R = AV - V\Gamma \tag{45}$$

we might reasonably expect it to be small. We have

$$V^{-1}AV = \Gamma + V^{-1}R = \Gamma + E, \tag{46}$$

where E is given by

$$VE = R. \tag{47}$$

The matrix $(\Gamma + E)$ is similar to A, and (for small R) approximately diagonal. We may hope, therefore, to be able to construct useful bounds for the eigenvalues of $\Gamma + E$ (and hence also of A). The evaluation of E – by calculating R from (45) and then solving equations (47) – should be carried out to a high order of accuracy, but provided equations (47) are not too ill-conditioned this can be done successfully. Notice that the diagonal elements of E provide correction terms for the eigenvalues. The following example illustrates the technique.

Consider the matrix

$$A = \begin{pmatrix} 2.05 & -1.05 & 1.045 \\ 0.95 & 0.05 & 1.045 \\ 0.97 & 3.03 & 1.30 \end{pmatrix} \tag{48}$$

Suppose that we have calculated the approximate eigenvalues $1, -1, 3$ with corresponding approximate eigenvectors

$$\begin{pmatrix} -3 \\ 1 \\ 4 \end{pmatrix}, \quad \begin{pmatrix} -1 \\ -1 \\ 2 \end{pmatrix}, \quad \begin{pmatrix} 1 \\ 1 \\ 2 \end{pmatrix}.$$

We find

$$\Gamma + E = \begin{pmatrix} 1.1 & 0 & 0 \\ 0.09 & -0.895 & 0.105 \\ 0.37 & 0.195 & 3.195 \end{pmatrix}.$$

We see that A has one eigenvalue exactly equal to 1.1, whilst the other two satisfy the inequalities

$$|d_2 + 0.895| \leqslant 0.195 \quad \text{and} \quad |d_3 - 3.195| \leqslant 0.565.$$

The values −0.895 and 3.195 are good approximations to the exact eigenvalues, −0.9 and 3.2, but the bounds obtained are rather wide. We can improve these bounds by the device employed in Section 7.3.2. We consider the matrix

$$\begin{pmatrix} 1.1 & 0 & 0 \\ 0.09\omega & -0.895 & 0.105\omega \\ 0.37 & 0.195/\omega & 3.195 \end{pmatrix}$$

choosing ω as small as possible consistent with keeping the Gerschgorin disc isolated. We may choose $\omega = 0.053$, which yields the bound

$$-0.905 \leqslant d_2 \leqslant -0.885$$

In the same way, the third eigenvalue is shown to satisfy

$$3.1875 \leqslant d_3 \leqslant 3.2025.$$

References

1. WILKINSON, J. H., 1965, *The Algebraic Eigenvalue Problem.* Oxford University Press.
2. ORTEGA, J. M., 1963, "An error analysis of Householder's method for the symmetric eigenvalue problem." *Numerische Math.* 5, 211–225.

APPENDIX

A SURVEY OF ESSENTIAL RESULTS FROM LINEAR ALGEBRA

8.1 Linear Vector Spaces

8.1.1

An ordered set $\{x_1, x_2, \ldots, x_n\}$ of n real (resp. complex) numbers is said to be a *vector*, $\mathbf{x}$, of dimension n; the numbers $x_1, x_2, \ldots, x_n$, are called the *co-ordinates*, or the *components*, of $\mathbf{x}$. Given any two vectors of dimension n, say

$$\mathbf{x} \equiv \{x_1, x_2, \ldots, x_n\} \quad \text{and} \quad \mathbf{y} \equiv \{y_1, y_2, \ldots, y_n\} \tag{1}$$

we define their *sum* as the vector (also of dimension n) given by

$$\mathbf{x} + \mathbf{y} \equiv \{x_1 + y_1,\ x_2 + y_2, \ldots, x_n + y_n\}.$$

Given any vector $\mathbf{x} = \{x_1, x_2, \ldots, x_n\}$ of dimension n, and any real (resp. complex) number λ (herinafter referred to as the *scalar* λ), we define the product of $\mathbf{x}$ and λ to be the vector given by

$$\lambda\mathbf{x} \equiv \{\lambda x_1, \lambda x_2, \ldots, \lambda x_n\}. \tag{2}$$

The set of all vectors of dimension n, together with the operations of vector addition and multiplication by scalars as defined above, is said to constitute the *linear space of all n-vectors*, or the *vector space of dimension n.*

8.1.2

It is easily confirmed that, for any given positive integer n, the linear space of all n-vectors satisfies the following conditions:

L1. Vector addition is associative, commutative, and distributive with respect to scalar multiplication:

$$\mathbf{x} + (\mathbf{y} + \mathbf{z}) = (\mathbf{x} + \mathbf{y}) + \mathbf{z}, \qquad \mathbf{x} + \mathbf{y} = \mathbf{y} + \mathbf{x}; \qquad \lambda(\mathbf{x} + \mathbf{y}) = \lambda\mathbf{x} + \lambda\mathbf{y}.$$

L2. $(\lambda + \mu)\mathbf{x} = \lambda\mathbf{x} + \mu\mathbf{x}; \quad \lambda(\mu\mathbf{x}) = (\lambda\mu)\mathbf{x}.$

L3. There is a unique vector, $\boldsymbol{\phi}$, called the *null vector* of dimension n, such that $\mathbf{x} + \boldsymbol{\phi} = \mathbf{x}$, for every $\mathbf{x}$.

L4. To each vector $\mathbf{x}$ there corresponds a unique inverse $(-\mathbf{x})$ which is such that $\mathbf{x} + (-\mathbf{x}) = \boldsymbol{\phi}$.

L5. $1\mathbf{x} = \mathbf{x}; \quad 0\mathbf{x} = \boldsymbol{\phi}.$

The null vector of dimension n is clearly that vector whose components are all zero, $\boldsymbol{\phi} \equiv \{0, 0, \ldots, 0\}$. In practice, there is little risk of confusion in using the ordinary numeral 0 to denote a null vector, irrespective of the dimension of the space with which we are concerned.

In general, any set of objects which is closed under some law of combination (called vector addition) and under a form of multiplication by real, or by complex, numbers (called scalar multiplication) is said to be a *linear* (or *vector*) space provided that the conditions L1–L5 are satisfied. When the scalar multipliers are confined to the real number system, the space is called a *real linear space*; otherwise it is called a *complex linear space.*

8.1.3 Linear Dependence

In what follows all vectors are assumed to have the same dimension, n. A vector $\mathbf{y}$ is said to be a *linear combination* of vectors $\mathbf{x}_1, \mathbf{x}_2, \ldots, \mathbf{x}_m$, or to *depend linearly on* those vectors, if there exist scalars $\lambda_1, \lambda_2, \ldots, \lambda_m$, such that

$$\mathbf{y} = \lambda_1\mathbf{x}_1 + \lambda_2\mathbf{x}_2 + \ldots + \lambda_m\mathbf{x}_m. \tag{3}$$

In particular, let $\mathbf{e}_k$ denote the n-vector whose k^{th} component is 1 and all of whose other components are 0. Then if $\mathbf{x} = (x_1, x_2, \ldots, x_n)$ is any vector of dimension n it can be expressed as a linear combination of the vectors $\mathbf{e}_1, \mathbf{e}_2, \ldots, \mathbf{e}_n$:–

$$\mathbf{x} = x_1\mathbf{e}_1 + x_2\mathbf{e}_2 + \ldots + x_n\mathbf{e}_n. \tag{4}$$

Thus, the space of all n-vectors is simply the set of all finite, linear combinations of the vectors $\mathbf{e}_k$, $1 \leqslant k \leqslant n$; this is usually expressed by saying that the vectors $\mathbf{e}_k$ *span* the space of all vectors of dimension n.

A set of vectors $\mathbf{x}_1, \mathbf{x}_2, \ldots, \mathbf{x}_m$ is said to be a *linearly dependent* set if and only if there exist scalars $\alpha_1, \alpha_2, \ldots, \alpha_m$, not all zero, such that

$$\alpha_1\mathbf{x}_1 + \alpha_2\mathbf{x}_2 + \ldots + \alpha_m\mathbf{x}_m = 0 \tag{5}$$

Otherwise the vectors are said to be *linearly independent.*

8.1.4

A set of linearly independent vectors which span a space is said to constitute a *basis* for that space. The vectors $\mathbf{e}_k$ discussed above are plainly linearly independent and hence form a basis for the space of all n-vectors. It can be shown that the number of vectors in any basis of a vector space is the same as that in any other basis. This fact allows us to give a formal definition of the term *dimension:*

A vector space E is said to be dimension n if there exists a basis for S which consists of precisely n vectors.

In view of this definition, it is clearly consistent to refer to the space of all n-vectors as a vector space of dimension n. Moreover, if E_1 and E_2 are any two vector spaces of the same dimension, n, then it is possible to set up a one-to-one correspondence between the members of E_1 and the members of E_2 such that:

if $\mathbf{x}_1$ and $\mathbf{y}_1$ in E_1 correspond to $\mathbf{x}_2$ and $\mathbf{y}_2$ in E_2 respectively and if λ and μ are any scalars, then

$$(\lambda \mathbf{x}_1 + \mu \mathbf{y}_1) \text{ corresponds to } (\lambda \mathbf{x}_2 + \mu \mathbf{y}_2).$$

Thus, the vector space structure of E_1 corresponds precisely with that of E_2; the two spaces are said to be *isomorphic*. It follows that every vector space of dimension n is isomorphic with the space of all n-vectors. For this reason the space of all n-vectors is often referred to as *the* vector space of dimension n (or, simply, as n-dimensional space).

8.1.5 Sub-spaces

A non-empty subset M of a linear space E is said to be a *subspace* of E if, whenever vectors $\mathbf{x}$ and $\mathbf{y}$ belong to M, so also does every linear combination $(\alpha\mathbf{x} + \beta\mathbf{y})$. Note that if $\mathbf{x}$ is any member of the subspace M then so also is the particular linear combination $\mathbf{x} - \mathbf{x} \equiv 0$. In fact, M is a subspace of E if and only if it constitutes a vector space in its own right, the null vector of which is the same as the null vector of E itself.

Example

Let E be the space of all vectors, $\mathbf{x} \equiv \{x_1, x_2, x_3\}$, of dimension 3. Then the subset of E comprising all vectors of the form $\{x_1, x_2, 0\}$ is a subspace of E – as indeed is any other plane which passes through the origin. Note that this subspace is isomorphic with the space of all vectors of dimension 2, for we can always set up the one-to-one correspondence

$$\{x_1, x_2, 0\} \longleftrightarrow \{x_1, x_2\}.$$

8.2 Linear Transformation

8.2.1

Let E and F be vector spaces of dimensions n and m respectively and let $\mathscr{A}$ denote a mapping of E into F which is such that

$$\text{(a)} \quad \mathscr{A}(\mathrm{x}+\mathrm{y}) = \mathscr{A}(\mathrm{x}) + \mathscr{A}(\mathrm{y}),$$

$$\text{(b)} \quad \mathscr{A}(\lambda \mathrm{x}) = \lambda \mathscr{A}(\mathrm{x}).$$

Such a mapping is called a *linear transformation* of E into F.

8.2.2

Let E_0 denote the set of all vectors $\mathbf{x}$ in E which are such that $\mathscr{A}(\mathbf{x}) = 0$. Every linear transformation must necessarily map the null vector of E into the null vector of F, i.e. E_0 always contains the null vector of E. Again, $\mathscr{A}$ will map any linear combination of vectors of E into the like linear combination of the corresponding vectors of F. It follows that E_0 is a certain vector subspace of E, called the *null space*, or *kernel*, of the linear transformation $\mathscr{A}$. The dimension of E_0 is called the *nullity* of the linear transformation.

Again, let F_1 denote the set of all vectors $\mathbf{y}$ in F which are such that $\mathscr{A}(\mathbf{x}) = \mathbf{y}$ for some vector $\mathbf{x}$ in E. (F_1 is often called the *image* of E under the linear transformation $\mathscr{A}$, or the *range* of $\mathscr{A}$.) In general, F_1 will be a subset of F but will not necessarily coincide with F. From considerations similar to those used above for E_0, it can be shown that F must actually be a vector subspace of F. The dimension of F_1 is called the *rank* of the linear transformation $\mathscr{A}$.

Example

Let E be the space of all n-vectors, F the space of all $(n-1)$ vectors, and $\mathscr{A}$ the mapping defined by $\mathscr{A}\{x_1, x_2, \ldots, x_n\} = \{x_2, x_3, \ldots, x_n\}$. Then F_1 coincides with F, so that $\mathscr{A}$ has rank $(n-1)$. On the other hand, a vector $\mathbf{x} = \{x_1, x_2, \ldots, x_n\}$ in E maps into the null vector of F if and only if $x_2 = x_3 = \ldots = x_n = 0$. Hence E_0 is spanned by the unit vector $\{1, 0, \ldots, 0\}$ and so has unit dimension, i.e. the nullity of $\mathscr{A}$ is 1.

In contrast, suppose that $\mathscr{A}$ maps *every* vector in E into the null vector of F. Then E_0 coincides with E and so $\mathscr{A}$ has nullity n; F_1 consists only of the null vector of F and so $\mathscr{A}$ has rank 0. This (rather trivial) example of a linear transformation of E into F is called the *zero transformation.*

8.2.3 Theorem

For any linear transformation $\mathscr{A}$ mapping E into F,

$$(\textit{nullity of } \mathscr{A}) + (\textit{rank of } \mathscr{A}) = \textit{dimension of } E. \tag{6}$$

Proof

Let E have dimension n, and suppose that $\mathscr{A}$ has nullity k and rank r. Choose a basis $\mathbf{e}_1, \mathbf{e}_2, \ldots, \mathbf{e}_k, \ldots, \mathbf{e}_n$ for E such that $(\mathbf{e}_1, \mathbf{e}_2, \ldots, \mathbf{e}_k)$ is a basis for E_0. If $\mathbf{x}$ is any vector in E then we can write

$$\mathbf{x} = \sum_{j=1}^{n} x_j \mathbf{e}_j,$$

and the image of this element of E with respect to $\mathscr{A}$ must be

$$\sum_{j=1}^{n} x_j \mathscr{A}(\mathbf{e}_j) = \sum_{j=k+1}^{n} x_j \mathscr{A}(\mathbf{e}_j)$$

Moreover, the vectors $\mathscr{A}(\mathbf{e}_j)$, $j = k+1, k+2, \ldots, n$, are linearly independent. For if not there exist scalars λ_j, not all zero, such that

$$\sum_{j=k+1}^{n} \lambda_j \mathscr{A}(\mathbf{e}_j) = \mathscr{A}\left[\sum_{j=k+1}^{n} \lambda_j \mathbf{e}_j\right] = 0.$$

Thus, $\sum_{j=k+1}^{n} \lambda_j \mathbf{e}_j$ lies in the null space of $\mathscr{A}$. But since $(\mathbf{e}_1, \ldots, \mathbf{e}_k)$ is a basis for E_0 there exist scalars μ_j such that

$$\sum_{j=1}^{k} \mu_j \mathbf{e}_j = \sum_{j=k+1}^{n} \lambda_j \mathbf{e}_j$$

which contradicts the fact that $\mathbf{e}_1, \mathbf{e}_2, \ldots, \mathbf{e}_n$, are linearly independent. Hence we must have $\lambda_j = 0$ for $j = k+1, k+2, \ldots, n$, and so the vectors $\mathscr{A}(\mathbf{e}_{k+1})$, $\mathscr{A}(\mathbf{e}_{k+2}), \ldots, \mathscr{A}(\mathbf{e}_n)$ form a basis for F_1, i.e. F_1 has dimension $n-k$, and so $r = n-k$.

8.2.4 Matrix Representation

Let the n-dimensional space E have basis $\{\mathbf{e}_1, \mathbf{e}_2, \ldots, \mathbf{e}_n\}$ and the m-dimensional space F have basis $\{\mathbf{f}_1, \mathbf{f}_2, \ldots, \mathbf{f}_m\}$. If $\mathscr{A}$ is a linear transformation of E into F then, for each j, the vector $\mathscr{A}(\mathbf{e}_j)$ is a linear combination of the basic vectors $\mathbf{f}_k$:

$$\mathscr{A}(\mathbf{e}_j) = a_{1j}\mathbf{f}_1 + a_{2j}\mathbf{f}_2 + \ldots + a_{mj}\mathbf{f}_j, \quad j = 1, 2, \ldots, n.$$

Now let $\mathbf{x}$ be any vector in E which is mapped into $\mathbf{y} = \mathscr{A}(\mathbf{x})$ in F. Since the $\mathbf{e}_k$ form a basis for E we must have

$$\mathbf{x} = x_1\mathbf{e}_1 + x_2\mathbf{e}_2 + \ldots + x_n\mathbf{e}_n,$$

and so,

$$\begin{aligned} \mathbf{y} &= \mathscr{A}(x_1\mathbf{e}_1 + x_2\mathbf{e}_2 + \ldots + x_n\mathbf{e}_n) \\ &= \sum_{i=1}^{m} (a_{i1}x_1 + a_{i2}x_2 + \ldots + a_{in}x_n)\,\mathbf{f}_i \end{aligned} \tag{7}$$

Hence, the linear transformation A mapping E into F can be represented by the $m \times n$ matrix

$$A \equiv [a_{ij}] = \begin{pmatrix} a_{11} & a_{12} & \dots & a_{1n} \\ a_{21} & a_{22} & \dots & a_{2n} \\ \dots & \dots & \dots & \dots \\ a_{m1} & a_{m2} & \dots & a_{mn} \end{pmatrix} \tag{8}$$

8.2.5 Matrix Algebra

The standard algebraic operations on matrices can be defined directly in terms of linear transformations.

(a) *Sum.* If $\mathscr{A}$ maps E into F and $\mathscr{B}$ also maps E into F then the sum of $\mathscr{A}$ and $\mathscr{B}$ is the linear transformation $(\mathscr{A} + \mathscr{B})$ given by,

$$(\mathscr{A} + \mathscr{B})(\mathbf{x}) \equiv \mathscr{A}(\mathbf{x}) + \mathscr{B}(\mathbf{x}), \qquad \text{for all } \mathbf{x} \text{ in } E.$$

If the matrix representations of $\mathscr{A}$ and $\mathscr{B}$ are $A \equiv [a_{ij}]$ and $B \equiv [b_{ij}]$ respectively, then it follows that $\mathscr{A} + \mathscr{B}$ has the matrix representation $[a_{ij} + b_{ij}]$.

(b) *Multiplication by Scalars.* If $\mathscr{A}$ maps E into F, and if λ is any scalar, then the linear transformation $\lambda\mathscr{A}$ is given by

$$(\lambda\mathscr{A})(\mathbf{x}) = \mathscr{A}(\lambda\mathbf{x}), \qquad \text{for all } \mathbf{x} \text{ in } E.$$

Accordingly, the matrix representation of $\lambda\mathscr{A}$ will be $[\lambda a_{ij}]$. It should be noted at this stage that the set of all linear transformations of E into F (and hence the set of all $m \times n$ matrices), equipped with the operations of addition and multiplication defined here, itself constitutes a vector space.

(c) *Product of Transformations.* Let $\mathscr{A}$ be a linear transformation which maps a space E, of dimension n, into a space F, of dimension m, and let $\mathscr{B}$ map F into a space G, of dimension p. The mapping $\mathscr{B}\mathscr{A}$, carrying E into G, is then defined to be the *composition* of the mappings $\mathscr{A}$ and $\mathscr{B}$ in the sense that

$$\mathscr{B}\mathscr{A}(\mathbf{x}) = \mathscr{B}\{\mathscr{A}(\mathbf{x})\}, \qquad \text{for all } \mathbf{x} \text{ in } E.$$

This results in the usual definition of matrix product; the matrix representation of the linear transformation $\mathscr{B}\mathscr{A}$ will be $BA \equiv [c_{ij}]$, where

$$c_{ij} = b_{i1}a_{1j} + b_{i2}a_{2j} + \dots + b_{im}a_{mj}, \qquad i = 1, 2, \dots, p.$$

(d) *The Identical Transformation.* Denote by $\mathscr{I}$ the linear transformation which maps E into itself in such a way that $\mathscr{I}(\mathbf{x}) = \mathbf{x}$ for every $\mathbf{x}$ in E. Then the matrix representation of $\mathscr{I}$ is easily seen to be $I \equiv [\delta_{ij}]$.

(e) *The Inverse Transformation.* Let $\mathscr{A}$ be a linear transformation which maps E into itself in a one-to-one manner. Then,

(i) $\mathscr{A}(\mathbf{x}) = 0$ if and only if $\mathbf{x} = 0$, i.e. the null space E_0 of $\mathscr{A}$ consists only of the null vector $\mathbf{x}$ of E, and so $\mathscr{A}$ has nullity 0;

(ii) if $\mathbf{y}$ is any vector in E then there exists just over one vector $\mathbf{x}$ in E such that $\mathbf{y} = \mathscr{A}(\mathbf{x})$. Thus, the range E_1 of $\mathscr{A}$ coincides with the vector space E itself, and so $\mathscr{A}$ has rank n;

(iii) there exists a one-to-one linear transformation of E into itself, called the inverse of $\mathscr{A}$ and written $\mathscr{A}^{-1}$, which is such that

$$\mathscr{A}\mathscr{A}^{-1} = \mathscr{A}^{-1}\mathscr{A} = \mathscr{I}.$$

Any transformation $\mathscr{A}$ for which the inverse $\mathscr{A}^{-1}$ exists is called a *non-singular transformation*; if $\mathscr{A}$ has no inverse then it is said to be a *singular* transformation. If a non-singular transformation $\mathscr{A}$ has matrix representation A, then the *inverse* transformation $\mathscr{A}^{-1}$ has matrix representation A^{-1}.

8.2.6 Row and Column Vectors

The matrix representation, A, of a linear transformation of n-dimensional space E into m-dimensional space F, as defined above, corresponds to the following algebraic formalism:

$$\mathbf{y}' = A\mathbf{x}.$$

That is to say, we are apparently committed to representing the members of E as column vectors (or $n \times 1$ matrices) and the members of F as row vectors (or $1 \times m$ matrices). In fact, this turns out to be no more than a harmless and convenient convention, since the vector space of all row vectors having m components is easily seen to be isomorphic with the space of all m-dimensional column vectors.

Usually, we shall take the "natural" representation of the n-vector $\{x_1, x_2, \ldots, x_n\}$ to be the column vector $\mathbf{x}$. However, it is worth noting that algebraic expressions involving matrices and row and column vectors can often be interpreted in different ways. For example, let $\mathbf{x}$ and $\mathbf{y}$ denote two arbitrary (real) n-vectors. The *scalar product*, $\mathbf{x}.\mathbf{y}$ or $\langle \mathbf{x}, \mathbf{y} \rangle$ may be defined as the number

$$x_1y_1 + x_2y_2 + \ldots + x_ny_n.$$

This may be written algebraically as the matrix product $\mathbf{x}'\mathbf{y}$, in which the n-vector $\{x_1, x_2, \ldots, x_n\}$ is now represented as the row vector $\mathbf{x}'$. On the other hand, recalling that $\mathbf{x}'$ is simply a $1 \times n$ matrix, we can always interpret the mapping $\mathbf{y} \rightarrow \mathbf{x}'\mathbf{y}$ as a linear transformation of n-dimensional space into one-dimensional space; the row vector $\mathbf{x}'$ this time is taken as a matrix representing a linear transformation. In what follows we shall revert to the ordinary notation $(x_1, \ldots, x_n)$ for both n-vectors and column vectors. The appropriate interpretation, where necessary, can always be made from the context.

8.3 Similarity Transformations

8.3.1

The matrix representation (7) of a linear transformation $\mathscr{A}$ pre-supposes a specific choice of basis for E and for F. A change of basis in either E or F will not, of course, affect the transformation $\mathscr{A}$ but will, in general, alter the matrix representation of $\mathscr{A}$. The point is best understood in terms of an example. Let E denote ordinary 3-space and consider two possible choices of basis for E:

Standard Basis	*Alternative Basis*
$\mathbf{e}_1 = (1, 0, 0)$	$\mathbf{e}_1' = (1, -1, 1)$
$\mathbf{e}_2 = (0, 1, 0)$	$\mathbf{e}_2' = (1, 2, 1)$
$\mathbf{e}_3 = (0, 0, 1)$	$\mathbf{e}_3' = (0, 2, 3)$

Let $\mathbf{x} \equiv (1, -6, 7)$. In terms of the standard basis, this means that

$$\mathbf{x} = \mathbf{e}_1 - 6\mathbf{e}_2 + 7\mathbf{e}_3 .$$

To express $\mathbf{x}$ in terms of the alternative basis, we need to determine numbers x_1', x_2', x_3', such that

$$\begin{pmatrix} 1 \\ -6 \\ 7 \end{pmatrix} = x_1'\mathbf{e}_1' + x_2'\mathbf{e}_2' + x_3'\mathbf{e}_3' = x_1' \begin{pmatrix} 1 \\ -1 \\ 1 \end{pmatrix} + x_2' \begin{pmatrix} 1 \\ 2 \\ 1 \end{pmatrix} + x_3' \begin{pmatrix} 0 \\ 2 \\ 3 \end{pmatrix}$$

i.e. such that

$$\begin{aligned} 1 &= x_1' + x_2' \\ -6 &= -x_1' + 2x_2' + 2x_3' \\ 7 &= x_1' + x_2' + 3x_3' \end{aligned}$$

which gives $x_1' = 4$, $x_2' = -3$, $x_3' = 2$, i.e. $\mathbf{x} = 4\mathbf{e}_1' - 3\mathbf{e}_2' + 2\mathbf{e}_3'$. To avoid confusion, we will denote x expressed in terms of the alternative basis by $[4, -3, 2]$.

Similarly, we let F denote ordinary 2-dimensional space with two possible choices of basis.

Standard Basis	*Alternative Basis*
$\mathbf{f}_1 = (1, 0)$	$\mathbf{f}_1' = (1, -1)$
$\mathbf{f}_2 = (0, 1)$	$\mathbf{f}_2' = (1, 1)$

We consider the linear transformation, $\mathscr{A}(x_1, x_2, x_3) = (x_3, x_1)$. To find the matrix representation of $\mathscr{A}$ with respect to the standard basis on E and the standard basis on F, we proceed as follows. First find the images under A of the basis vectors $\mathbf{e}_1$, $\mathbf{e}_2$, and $\mathbf{e}_3$:

$$\begin{aligned}\mathbf{e}_1 &\equiv (1,0,0) \to (0,1) = 0\mathbf{f}_1 + 1\mathbf{f}_2\\ \mathbf{e}_2 &\equiv (0,1,0) \to (0,0) = 0\mathbf{f}_1 + 0\mathbf{f}_2\\ \mathbf{e}_3 &\equiv (0,0,1) \to (1,0) = 1\mathbf{f}_1 + 0\mathbf{f}_2\end{aligned}$$

The required matrix, A_1, is therefore $A_1 = \begin{pmatrix} 0 & 0 & 1 \\ 1 & 0 & 0 \end{pmatrix}$.

As a check, we can find the image of the vector $\mathbf{x} = (1, -6, 7)$.

$$A_1\mathbf{x} = \begin{pmatrix} 0 & 0 & 1 \\ 1 & 0 & 0 \end{pmatrix} \begin{pmatrix} 1 \\ -6 \\ 7 \end{pmatrix} = \begin{pmatrix} 7 \\ 1 \end{pmatrix},$$

which is consistent with the mapping $\mathscr{A}$.

To find the matrix representation, A_2, of $\mathscr{A}$ with respect to the basis $\mathbf{e}'_k$ on E and the basis $\mathbf{f}'_k$ on F:

$$\begin{aligned}\mathbf{e}'_1 &\equiv (1,-1,1) \to (1,1) = 0\mathbf{f}'_1 + 1\mathbf{f}'_2\\ \mathbf{e}'_2 &\equiv (1,2,1) \to (1,1) = 0\mathbf{f}'_1 + 1\mathbf{f}'_2\\ \mathbf{e}'_3 &\equiv (0,2,3) \to (3,0) = \tfrac{3}{2}\mathbf{f}'_1 + \tfrac{3}{2}\mathbf{f}'_2.\end{aligned}$$

Hence,

$$A_2 = \begin{pmatrix} 0 & 0 & \frac{3}{2} \\ 1 & 1 & \frac{3}{2} \end{pmatrix}$$

Once again, we can check the result by finding the image of the vector $\mathbf{x} \equiv (1, -6, 7)$. Recall that in terms of the alternative basis vectors $\mathbf{e}'_k$ on E we have $\mathbf{x} \equiv [4, -3, 2]$. Then,

$$A_2\mathbf{x} = \begin{pmatrix} 0 & 0 & \frac{3}{2} \\ 1 & 1 & \frac{3}{2} \end{pmatrix} \begin{pmatrix} 4 \\ -3 \\ 2 \end{pmatrix} = \begin{pmatrix} 3 \\ 4 \end{pmatrix}.$$

Finally, we have $[3, 4] = 3\mathbf{f}'_1 + 4\mathbf{f}'_2 = (7, 1)$, in the standard basis on F.

8.3.2

The identity transformation, $\mathscr{I}$, on E carries every vector into itself: $\mathscr{I}(\mathbf{x}) = \mathbf{x}$ for every $\mathbf{x}$. We can regard this, if we wish, as a transformation from the space E, with basis $\mathbf{e}'_k$, to the space E, with basis $\mathbf{e}_k$. As such, it will have a certain matrix representation, $I_{\mathbf{e}'\mathbf{e}}$, which we can find in the case of the present example as follows:

$$\begin{aligned} \mathbf{e}_1' &= 1\mathbf{e}_1 - 1\mathbf{e}_2' + 1\mathbf{e}_3 \\ \mathbf{e}_2' &= 1\mathbf{e}_1 + 2\mathbf{e}_2 + 1\mathbf{e}_3 \\ \mathbf{e}_3' &= 0\mathbf{e}_1 + 2\mathbf{e}_2 + 3\mathbf{e}_3 \end{aligned} \qquad \text{so that } I_{\mathbf{e}'\mathbf{e}} = \begin{pmatrix} 1 & 1 & 0 \\ -1 & 2 & 2 \\ 1 & 1 & 3 \end{pmatrix}$$

On the other hand, we could equally well regard it as a transformation from the space E, with basis $\mathbf{e}_k$, to the space E, with basis $\mathbf{e}_k'$. The matrix representation this time would be

$$I_{\mathbf{ee}'} = \begin{pmatrix} \frac{4}{9} & -\frac{1}{3} & \frac{2}{9} \\ \frac{5}{9} & \frac{1}{3} & -\frac{2}{9} \\ -\frac{1}{3} & 0 & \frac{1}{3} \end{pmatrix} = I_{\mathbf{e}'\mathbf{e}}^{-1}$$

Similarly on F we have $\mathbf{f}_1' = 1\mathbf{f}_1 - 1\mathbf{f}_2$, $\mathbf{f}_2' = 1\mathbf{f}_1 + 1\mathbf{f}_2$ so that $I_{\mathbf{f}'\mathbf{f}} = \begin{pmatrix} 1 & 1 \\ -1 & 1 \end{pmatrix}$

and

$$I_{\mathbf{ff}'} = \begin{pmatrix} 1 & 1 \\ -1 & 1 \end{pmatrix}^{-1} = \begin{pmatrix} \frac{1}{2} & -\frac{1}{2} \\ \frac{1}{2} & \frac{1}{2} \end{pmatrix}.$$

In order to map $\mathbf{x}$ into its image $\mathscr{A}(\mathbf{x})$, using the matrix representation A_2 of $\mathscr{A}$, the following steps are needed:

(i) convert from the $\mathbf{e}_k$ to the $\mathbf{e}_k'$:– $\mathbf{x} \to I_{\mathbf{ee}'}\mathbf{x}$,

(ii) multiply by A_2:– $A_2 I_{\mathbf{ee}'}\mathbf{x}$,

(iii) convert from the $\mathbf{f}_k'$ to the $\mathbf{f}_k$:– $I_{\mathbf{f}'\mathbf{f}} A I_{\mathbf{ee}'}\mathbf{x}$.

It follows that

$$A_1 = I_{\mathbf{f}'\mathbf{f}} A_2 I_{\mathbf{ee}'} = \begin{pmatrix} 1 & 1 \\ -1 & 1 \end{pmatrix} \begin{pmatrix} 0 & 0 & \frac{3}{2} \\ 1 & 1 & \frac{3}{2} \end{pmatrix} \begin{pmatrix} \frac{4}{9} & -\frac{1}{3} & \frac{2}{9} \\ \frac{5}{9} & \frac{1}{3} & -\frac{2}{9} \\ -\frac{1}{3} & 0 & \frac{1}{3} \end{pmatrix}$$

Equivalently,

$$A_2 = I_{\mathbf{ff}'} A_1 I_{\mathbf{e}'\mathbf{e}} = \begin{pmatrix} \frac{1}{2} & -\frac{1}{2} \\ \frac{1}{2} & \frac{1}{2} \end{pmatrix} \begin{pmatrix} 0 & 0 & 1 \\ 1 & 0 & 0 \end{pmatrix} \begin{pmatrix} 1 & 1 & 0 \\ -1 & 2 & 2 \\ 1 & 1 & 3 \end{pmatrix}$$

8.3.3 Similar Matrices

Let T denote a linear transformation of an n-dimensional vector space E into itself. If E has a basis $(\mathbf{e}_k)$, $1 \leqslant k \leqslant n$, then there will be an $n \times n$ matrix, A, which represents T with respect to this basis. Again, if E has another, distinct, basis $(\mathbf{e}'_k)$, $1 \leqslant k \leqslant n$, then there will be a different matrix representation, B, of the linear transformation T with respect to $(\mathbf{e}'_k)$. Using arguments similar to those of the preceding discussion, we can show that

$$B = I_{\mathbf{ee}'} A I_{\mathbf{e}'\mathbf{e}} = I_{\mathbf{e}'\mathbf{e}}^{-1} A I_{\mathbf{e}'\mathbf{e}} \tag{9}$$

where, as before, $I_{\mathbf{e}'\mathbf{e}}$ denotes the matrix representation of the identity transformation on E with respect to the bases $\mathbf{e}'$ and $\mathbf{e}$. Matrices such as A and B which represent the same linear transformation with respect to different bases are said to be *similar*. A necessary and sufficient condition for similarity is that there exist a non-singular matrix H such that

$$B = H^{-1} A H. \tag{10}$$

8.3.4 Rank

When there is no risk of confusion we will use the same symbol A to denote both a linear transformation and its matrix representation. We suppose that A maps n-dimensional space E into m-dimensional space F, and that E and F have the (fixed) bases $(\mathbf{e}_1, \mathbf{e}_2, \ldots, \mathbf{e}_n)$ and $(\mathbf{f}_1, \mathbf{f}_2, \ldots, \mathbf{f}_m)$ respectively. For the moment, we adopt the following notation for the rows and columns of the matrix $A = [a_{ij}]$: $A_i \equiv (a_{i1}, a_{i2}, \ldots, a_{in})$; $A^j \equiv (a_{1j}, a_{2j}, \ldots, a_{mn})'$. The image of the base vector $\mathbf{e}_j$ is

$$A\mathbf{e}_j = \sum_{i=1}^{m} a_{ij}\, \mathbf{f}_i \quad j = 1, 2, \ldots, n \tag{11}$$

i.e. the coordinates of $A\mathbf{e}_j$ with respect to the $\mathbf{f}_i$ are given by the column vector A^j. Now consider the range, F_1, of the transformation A; any vector in F_1 is a linear combination of the vectors $A\mathbf{e}_j$, $1 \leqslant j \leqslant n$, so that F_1 is spanned by this set of vectors. The rank of A is the dimension of F_1, and this is just the largest set of linearly independent vectors taken from $A\mathbf{e}_1, A\mathbf{e}_2, \ldots, A\mathbf{e}_n$. Hence, it must be equal to the largest number of linearly independent columns of $[a_{ij}]$. This number is usually called the *column rank* of the matrix $[a_{ij}]$, and the space spanned by the linearly independent columns is called the *column space* of the matrix.

Let $A \equiv (a_{ij})$ have column rank r, and suppose that the r linearly independent columns are $B^1, B^2, \ldots, B^r$.

Then, for any column A^j we can write

$$A^j = \sum_{i=1}^{r} \alpha_{ij} B^i \qquad j = 1, 2, \ldots, n$$

so that

$$a_{kj} = \sum_{i=1}^{r} \alpha_{ij}\, b_{ki} \qquad \begin{array}{l} j = 1, 2, \ldots, n \\ k = 1, 2, \ldots, m \end{array}$$

where b_{ki} is the kth element of the column B^i. This can be re-written in terms of the rows A_i to give,

$$A_k = \sum_{i=1}^{r} b_{ki}\, \alpha_i$$

where α_i is the ith row of the $r \times n$ matrix $[\alpha_{ij}]$. If we define *row space* and *row rank* of $[a_{ij}]$ in an analogous fashion to column space and column rank, it follows from the above result that the row space is spanned by the set of r n-vectors α_i. Hence its dimension is less than or equal to r. If the above reasoning is applied in exactly the same way to the transpose of $[a_{ij}]$, we can show equally well that the dimension of the column space is less than or equal to that of the row space. Hence, the row rank and the column rank must be equal. In other words:

The number of linearly independent rows of any matrix is equal to the number of linearly independent columns.

8.3.5 Eigenvalues and Eigenvectors

If A is a linear transformation mapping x into itself, then any non-zero vector X is said to be an *eigenvector* of A if

$$A(\mathbf{x}) = \lambda \mathbf{x} \tag{12}$$

for some scalar λ; the scalar λ itself is called an *eigenvalue* of A. Let E have basis $\mathbf{e} \equiv (\mathbf{e}_1, \mathbf{e}_2, \ldots, \mathbf{e}_n)$ and let A_1 be the matrix representation of A with respect to this basis. If each basis vector $\mathbf{e}_k$ is an eigenvector of A then,

$$A(\mathbf{e}_k) = \lambda_k \mathbf{e}_k \qquad k = 1, 2, \ldots, n.$$

Hence the matrix A_1 will be the diagonal matrix,

$$A_1 = \text{diag}(\lambda_i).$$

On the other hand, if $A_1 = \text{diag}(\lambda_i)$ then,

$$A(\mathbf{e}_j) = \sum_{i=1}^{n} a_{ij}\mathbf{e}_j = \lambda_j \mathbf{e}_j, \qquad j = 1, 2, \ldots, n.$$

Hence, the matrix representation A_1 is diagonal if and only if the basis $\mathbf{e}_k$, $1 \leqslant k \leqslant n$, consists entirely of eigenvectors of A.

Now suppose that

$$\mathbf{y} = \sum_{i=1}^{n} y_i \mathbf{e}_i$$

is an eigenvector of the linear transformation A. Then:–

$$A(\mathbf{y}) = \lambda \mathbf{y} \qquad \text{and} \qquad A_1 \mathbf{y} = \lambda \mathbf{y},$$

where $\mathbf{y}$ denotes the column vector $(y_1, y_2, \ldots, y_n)'$. The matrix equation can be re-written in the form,

$$(\lambda I_n - A_1)\mathbf{y} = 0$$

which shows that y lies in the null space of the matrix $(\lambda I_n - A_1)$. This matrix is called the *characteristic matrix* of A. The matrix equation is equivalent to a system of homogeneous linear equations which has a non-zero solution if and only if,

$$|\lambda I_n - A_1| = 0. \tag{13}$$

This is the *characteristic equation* of the matrix A_1. The determinant can be expanded as a polynomial in λ, called the *characteristic polynomial* of A_1. The zeros of this polynomial are precisely the eigenvalues of A_1.

8.4 Vector and Matrix Norms

8.4.1

In ordinary 2-dimensional or 3-dimensional space there is an obvious and natural sense in which one would use the term "length of a vector". Thus, if $\mathbf{x} \equiv (x_1, x_2, x_3)$ then we would normally understand the length of $\mathbf{x}$ to be the number,

$$\mathbf{x} \equiv +\sqrt{\{x_1^2 + x_2^2 + x_2^2\}}.$$

Conceptually, this is just the same as the "distance from the origin" of the point whose coordinates are x_1, x_2, and x_3. As a result, we go on to speak of the distance between two points, say (x_1, x_2, x_3) and (y_1, y_2, y_3), as the length of the vector $\mathbf{x} - \mathbf{y}$:–

$$d(\mathbf{x}, \mathbf{y}) \equiv \|\mathbf{x} - \mathbf{y}\| = +\sqrt{\{(x_1 - y_1)^2 + (x_2 - y_2)^2 + (x_3 - y_3)^2\}}.$$

The concept of distance allows us to talk meaningfully about convergence and limiting processes in space, since there will be a specific sense in which we can speak of a sequence of points (or vectors) approaching more and more closely to some particular limit point. We wish to generalise these ideas to n-dimensional space (and, in fact, to vector spaces of any kind) and to do so it is necessary to isolate the essential formal properties of "length" and of "distance".

8.4.2 Norms

A norm is a real-valued function, $\|\mathbf{x}\|$, defined on a vector space, E, which satisfies the following conditions:

N1. $\|\mathbf{x}\| \geqslant 0$ for every $\mathbf{x}$ in E, and $\|\mathbf{x}\| = 0$ if and only if $\mathbf{x}$ is the null vector of E.

N2. $\|\mathbf{x} + \mathbf{y}\| \leqslant \|\mathbf{x}\| + \|\mathbf{y}\|$. (Triangle Inequality)

N3. $\|\alpha\mathbf{x}\| = |\alpha|\ \|\mathbf{x}\|$, for any scalar α.

A vector space E, together with a norm $\|\mathbf{x}\|$ defined on E, is usually referred to as a *normed linear space.* The *distance*, $d(\mathbf{x}, \mathbf{y})$, between any two members $\mathbf{x}$ and $\mathbf{y}$ of E is then well defined as,

$$d(\mathbf{x} - \mathbf{y}) \equiv \|\mathbf{x} - \mathbf{y}\| \tag{14}$$

and the real-valued function $d(\mathbf{x}, \mathbf{y})$ is often spoken of as a *metric* on E. Finally, a sequence of vectors $\mathbf{x}^{(1)}, \mathbf{x}^{(2)}, \mathbf{x}^{(3)}, \ldots$, in E is said to converge to $\mathbf{x}$ in E as its limit if and only if,

$$\lim_{k\to\infty} d(\mathbf{x}, \mathbf{x}^{(k)}) \equiv \lim_{k\to\infty} \|\mathbf{x} - \mathbf{x}^{(k)}\| = 0 \tag{15}$$

Convergence in this mode is described as convergence "in the sense of the norm $\|\mathbf{x}\|$ on E" (more briefly as "convergence in norm").

8.4.3 Special Norms

The straightforward generalisation of the usual concept of distance in 3-dimensional space leads to the definition of the so-called *Euclidean norm* in n-dimensional space:

$$\|\mathbf{x}\|_2 = +\sqrt{\{|x_1|^2 + |x_2|^2 + \ldots + |x_n|^2\}} = \left[\sum_{r=1}^{n} |x_r|^2\right]^{\frac{1}{2}}. \tag{16}$$

It is often convenient to use other definitions of norm in n-dimensional space. A very useful sequence of possible norms (including the Euclidean norm as a special case) can be obtained by writing:

$$\|\mathbf{x}\|_p = \left[\sum_{r=1}^{n} |x_r|^p\right]^{1/p} \qquad p = 1, 2, 3, \ldots \tag{17}$$

Moreover, if we allow p to tend to ∞ it can be shown that

$$\lim_{p\to\infty} \left[\sum_{r=1}^{n} |x_r|^p\right]^{1/p} = \max_{1 \leqslant r \leqslant n} |x_r|. \tag{18}$$

It is easily confirmed that this limiting value satisfies the conditions N1.–N3. for a norm on n-dimensional space and it is usual to write,

$$\|\mathbf{x}\|_\infty = \max_{1 \leqslant r \leqslant n} |x_r|. \tag{19}$$

In the case of the norm $\|\mathbf{x}\|_p$, for $p \geqslant 1$, the "triangle inequality" N2. is simply the statement of a classical result in Analysis known as *Minkowski's Inequality:*

$$\left[\sum_{r=1}^{n} |x_r + y_r|^p\right]^{1/p} \leqslant \left[\sum_{r=1}^{n} |x_r|^p\right]^{1/p} + \left[\sum_{r=1}^{n} |y_r|^p\right]^{1/p}.$$

In the present context, the norms of particular practical interest are:

$$\|\mathbf{x}\|_1 = |x_1| + |x_2| + \ldots + |x_n| = \left\{\sum_{r=1}^{n} |x_r|\right\} \tag{20}$$

$$\|\mathbf{x}\|_2 = +\sqrt{\{|x_1|^2 + |x_2|^2 + \ldots + |x_n|^2\}} = \left\{\sum_{r=1}^{n} |x_r|^2\right\}^{\frac{1}{2}} \tag{21}$$

$$\|\mathbf{x}\|_\infty = \max |x_r|, \qquad 1 \leqslant r \leqslant n. \tag{22}$$

Once a specific norm has been chosen for n-dimensional space, there will be a corresponding concept of *distance:*

$$d_p(\mathbf{x},\mathbf{y}) = \|\mathbf{x}-\mathbf{y}\|_p = \left[\sum_{r=1}^{n} |x_r - y_r|^p\right]^{1/p}$$

$$d_\infty(\mathbf{x},\mathbf{y}) = \max \|x_r - y_r|, \quad 1 \leqslant r \leqslant n.$$

Accordingly, one would expect to find a corresponding mode of convergence peculiar to each specific choice of p:

The sequence $\{\mathbf{x}^{(k)}\}$ in E converges to the limit $\mathbf{x}$ in the sense of the norm $\|\mathbf{x}\|$ on E if and only if,

$$\lim_{k\to\infty} \|\mathbf{x} - \mathbf{x}^{(k)}\|_p = 0.$$

In fact, however, things are rather simpler than this; if the sequence $\{\mathbf{x}^{(k)}\}$ converges to the limit $\mathbf{x}$ in the sense of any of the norms described above, then we must have

$$\lim_{k\to\infty} x_1^{(k)} = x_1$$

$$\lim_{k\to\infty} x_2^{(k)} = x_2$$

$$\cdots\cdots\cdots\cdots$$

$$\lim_{k\to\infty} x_n^{(k)} = x_n,$$

i.e. each co-ordinate of the limit vector $\mathbf{x}$ is the limit (in the ordinary sense) of the sequence of the corresponding coordinates of the vectors $\mathbf{x}^{(k)}$.

8.4.4 Norm of a Bounded Transformation

Quite generally, let A be a linear transformation mapping the elements of one normed linear space, E, into those of another, F. Then A is said to be *continuous* with respect to the norms on E and F if the following situation obtains:

Let $\{\mathbf{x}^{(k)}\}$ be any sequence of vectors in E which converges to the limit $\mathbf{x}$ in the sense of the norm on E; then the corresponding sequence $\{A(\mathbf{x}^{(k)})\}$ of the image vectors in F converges to the limit $A(\mathbf{x})$ in the sense of the norm on F.

A is a continuous linear transformation of E into F if and only if it is *bounded* in the sense that there exists a finite positive number M such that,

$$\|A(\mathbf{x})\| \leqslant M\|\mathbf{x}\| \tag{23}$$

for every $\mathbf{x}$ in E. The smallest number M such that this relation is true for all $\mathbf{x}$ in E is called the *norm of the bounded linear transformation* A, and is written as $\|A\|$. It can be defined explicitly by:

$$\|A\| = \max_{\|\mathbf{x}\|\neq 0} \frac{\|A(\mathbf{x})\|}{\|\mathbf{x}\|}$$

or equivalently by,

$$\|A\| = \max_{\|\mathbf{x}\|=1} \|A(\mathbf{x})\|. \tag{24}$$

The term "norm" is justified since, as described in Section 9.2.5, there is a natural way to define the sum of two linear transformations and to define the product of a linear transformation with a scalar. Equipped with these two operations, the set of all linear transformations mapping E into F is seen to constitute a vector space in its own right. Moreover, the quantity $\|A\|$ defined above actually satisfies the properties N1.–N3. which characterise a norm on that space. Note that the fundamental property of a bounded linear transformation can be expressed in the form,

$$\|A(\mathbf{x})\| \leqslant \|A\| \; \|\mathbf{x}\|, \quad \text{for all } \mathbf{x} \text{ in } E.$$

8.4.5 Matrix Norms

Let A now denote any $n \times n$ square matrix. With the usual definitions of matrix addition and scalar multiplication, the set of all such matrices forms a vector space. A real-valued function, $\|A\|$, defined on this space will constitute a norm if the formal rules N1.–N3. described in Section 9.4.3 are satisfied. That is to say we should have,

MN1. $\|A\| \geqslant 0$ always, and $\|A\| = 0$ if and only if A is the null matrix.
MN2. $\|A + B\| \leqslant \|A\| + \|B\|$.
MN3. $\|\alpha A\| = |\alpha| \; \|A\|$, for every scalar α.

Any such real-valued function will be called a *matrix norm*. However, as is easy to confirm, every matrix A, with n rows and n columns, represents a certain linear transformation of the space, E, of all n-vectors into itself. (We assume for the moment that some particular basis has been fixed on for E.) Now let $\|\mathbf{x}\|$ be some specific (vector) norm on E. If $\|A\|$ is any given matrix norm which is such that

$$\|A(\mathbf{x})\| \leqslant \|A\| \; \|\mathbf{x}\| \tag{25}$$

for every vector $\mathbf{x}$ in E and for every $n \times n$ matrix A, then the matrix norm $\|A\|$ is said to be *consistent* (or *compatible*) with the vector norm $\|\mathbf{x}\|$.

In particular, the matrix norm defined by

$$\|A\| = \max_{\|\mathbf{x}\|=1} \|A(\mathbf{x})\|.$$

is obviously consistent with $\|\mathbf{x}\|$, and is actually said to be *subordinate to* $\|\mathbf{x}\|$. Note that for such a norm the unit matrix always has unit norm; again, a subordinate norm is always less than or equal to any other matrix norm which is consistent with the same vector norm.

A necessary condition for a matrix norm to be consistent with a given vector norm $\|\mathbf{x}\|$ is obtained when we consider matrix multiplication. Thus, suppose that the matrix norm $\|A\|$ is consistent with the vector norm $\|\mathbf{x}\|$, and consider, the product $C = AB$:

$$\|C(\mathbf{x})\| = \|AB(\mathbf{x})\| \leqslant \|A\| \; \|B(\mathbf{x})\| \leqslant \|A\| \; \|B\| \; \|\mathbf{x}\|$$

i.e. for arbitrary matrices A, B, we must have

MN4. $\|AB\| \leqslant \|A\| \; \|B\|$.

8.4.6 Some Special Matrix Norms

If the vector norm on E is $\|\mathbf{x}\|_\infty$ then for the corresponding subordinate norm $\|A\|_\infty$ we have:

$$\|A\mathbf{x}\|_\infty = \max_i \left| \sum_{k=1}^{n} a_{ik} x_k \right| \leqslant \max_i \sum_{k=1}^{n} |a_{ik}| \, |x_k|.$$

If $\max_i \sum_{k=1}^{n} |a_{ik}|$ is reached when $i = \alpha$, take for $\mathbf{x}$ the vector defined by, $x_k = |a_{\alpha k}|/a_{\alpha k}$, if $a_{\alpha k} \neq 0$, and $x_k = 1$ otherwise. Then $\|\mathbf{x}\|_\infty = 1$ and we have

$$\|A\|_\infty = \sup_{\|\mathbf{x}\|_\infty = 1} \|A\mathbf{x}\|_\infty = \max_i \sum_{k=1}^{n} |a_{ik}|. \tag{27}$$

Similarly, if the vector norm on E is $\|\mathbf{x}\|_1$ then we have

$$\|A\mathbf{x}\|_1 = \sum_{i=1}^{n} \left| \sum_{k=i}^{n} a_{ik} x_k \right| \leqslant \sum_{i=1}^{n} \sum_{k=1}^{n} |a_{ik}| \, |x_k| \leqslant \sum_{k=1}^{n} |x_k| \max_k \sum_{i=1}^{n} |a_{ik}|.$$

If $\max_k \sum_{i=1}^{n} |a_{ik}|$ is attained when $k = \beta$, take for $\mathbf{x}$ the vector defined by,

$$x_k = 0 \text{ if } k \neq \beta, \quad \text{and } x_\beta = 1.$$

Then $\|\mathbf{x}\|_1 = 1$ and we have

$$\|A\|_1 = \sup_{\|\mathbf{x}\|_1 = 1} \|A\mathbf{x}\|_1 = \max_k \sum_{i=1}^{n} |a_{ik}|. \tag{28}$$

Finally, let the vector norm on E be given by

$$\|\mathbf{x}\|_2 = \left[\sum_{k=1}^{n} |x_k|^2 \right]^{\frac{1}{2}}$$

Then $\|\mathbf{x}\|^2 = \bar{\mathbf{x}}'\mathbf{x}$ and $\|A\mathbf{x}\|_2^2 = (\overline{A\mathbf{x}})'(A\mathbf{x}) = \bar{\mathbf{x}}'\bar{A}'A\mathbf{x}$. Hence, the subordinate matrix norm $\|A\|_2$ is given by

$$\|A\|_2^2 = \max_{\mathbf{x} \neq 0} \frac{\|A\mathbf{x}\|_2^2}{\|\mathbf{x}\|_2^2} = \max \frac{\bar{\mathbf{x}}'\bar{A}'A\mathbf{x}}{\bar{\mathbf{x}}'\mathbf{x}}.$$

Now the matrix $\bar{A}'A$ is symmetric and positive semi-definite (i.e. for all $\mathbf{x}$ we must have $\bar{\mathbf{x}}'\bar{A}'A\mathbf{x} \geqslant 0$). It follows that the eigenvalues of this matrix are all real and non-negative. If these eigenvalues are denoted by σ_k^2, $1 \leqslant k \leqslant n$, where $\sigma_1^2 \geqslant \sigma_2^2 \geqslant \ldots \geqslant \sigma_n^2 \geqslant 0$, and if the corresponding (orthonormalised) real eigenvectors are $\mathbf{x}_1, \mathbf{x}_2, \ldots, \mathbf{x}_n$, then for any non-zero vector $\mathbf{x}$ we can write

$\mathbf{x} = \sum_{k=1}^{n} c_k \mathbf{x}_k$, and so

$$\frac{\|A\mathbf{x}\|_2^2}{\|\mathbf{x}\|_2^2} = \frac{\mathbf{x}'(\bar{A}'A\mathbf{x})}{\bar{\mathbf{x}}'\mathbf{x}} = \left[\sum_{k=1}^{n} |c_k|^2 \sigma_k^2\right] \Big/ \left[\sum_{k=1}^{n} |c_k|^2\right].$$

Hence,

$$0 \leqslant \sigma_n^2 \leqslant \left\{\frac{\|A\mathbf{x}\|_2^2}{\|\mathbf{x}\|_2^2}\right\} \leqslant \sigma_1^2$$

Further, by choosing $\mathbf{x} = \mathbf{x}_n$, it is clear that the value σ_1^2 can be attained. That is to say

$$\|A\|_2 = \max_{\mathbf{x} \neq \mathbf{o}} \left[\frac{\|A\mathbf{x}\|_2^2}{\|\mathbf{x}\|_2^2}\right]^{\frac{1}{2}} = \sigma_1. \tag{29}$$

This is often referred to as the *spectral norm* of the matrix A.

8.4.7 The Euclidean Norm

In addition to the subordinate norms discussed above, another important matrix norm is the so-called *Euclidean* norm:

$$\|A\|_E = \left[\sum_{i=1}^{n} \sum_{j=1}^{n} |a_{ij}|^2\right]^{\frac{1}{2}} \tag{30}$$

The term "Euclidean" derives from the fact that $\|A\|_E$ is actually the same as the ordinary Euclidean vector norm of a vector of dimension n^2 whose co-ordinates are the elements a_{ij} of the matrix A. Nevertheless, the term is confusing since $\|A\|_E$ is not the subordinate norm for the Euclidean vector norm (except trivially in the case $n = 1$) although it is always consistent with it. For, using the Cauchy-Schwarz inequality, we have

$$\left|\sum_{k=1}^{n} a_{ik} x_k\right|^2 \leqslant \sum_{k=1}^{n} |a_{ik}|^2 \sum_{k=1}^{n} |x_k|^2, \quad \text{and so}$$

$$\|A\mathbf{x}\|_2^2 = \sum_{i=1}^{n} \left|\sum_{k=1}^{n} a_{ik} x_k\right|^2 \leqslant \sum_{i=1}^{n} \left\{\sum_{k=1}^{n} |a_{ik}|^2\right\} \sum_{k=1}^{n} |x_k|^2$$

$$= \left[\sum_{i=1}^{n} \sum_{k=1}^{n} |a_{ik}|^2\right] \sum_{k=1}^{n} |x_k|^2 = \|A\|_E^2 \, \|\mathbf{x}\|_2^2,$$

so that $\|A\mathbf{x}\|_2 \leqslant \|A\|_E \, \|\mathbf{x}\|_2$. (consistency)

On the other hand, $\|A\|_E$ clearly cannot be subordinate with respect to any vector norm for $n > 1$, since for the unit matrix I we are bound to get

$$\|I\|_E = +\sqrt{n}.$$

8.4.8 Rectangular Matrices and Norms

Any $m \times n$ matrix can be regarded as a representation of a linear transformation mapping n-dimensional space into m-dimensional space. Hence there is no need to restrict the material on matrix norms given in Sections 9.4.5. and 9.4.6 to square matrices. In point of fact, such a general treatment of matrix forms is not particularly useful except perhaps in the case of row and column vectors:

For a column vector $\mathbf{a}$ it is easy to confirm that the norms $\|\mathbf{a}\|_1$, $\|\mathbf{a}\|_\infty$, $\|\mathbf{a}\|_2$, mean the same thing whether they are defined as vector norms or as the appropriate $(m \times 1)$ matrix norms. Evaluating the matrix norms for row vectors we find that,

$$\|\mathbf{a}'\|_1 = \|\mathbf{a}\|_\infty, \qquad \|\mathbf{a}'\|_\infty = \|\mathbf{a}\|_1, \qquad \|\mathbf{a}'\|_2 = \|\mathbf{a}\|_2. \tag{31}$$

The first two results in (31) are obvious. The third follows from the fact that, in general, the matrices AA' and $A'A$ have the same eigenvalues. (For, let λ be an eigenvalue of AA' and $\mathbf{u}$ a corresponding eigenvector.

Then $$AA'\mathbf{u} = \lambda\mathbf{u}, \quad \text{and so} \quad \mathbf{u}'A'A = \lambda\mathbf{u}'.$$

Hence, $$\mathbf{u}'A'A\mathbf{u} = \lambda\mathbf{u}'\mathbf{u} = \mathbf{u}'(\lambda\mathbf{u}) = \mathbf{u}'(AA'\mathbf{u}).$$

Thus, $$A'A\mathbf{u} = AA'\mathbf{u} = \lambda\mathbf{u}$$

and so λ must be an eigenvalue of $A'A$.)

As an example of the way in which norms of row and column vectors might be manipulated, consider an expression of the form $\mathbf{x}'A\mathbf{y}$. Writing this as $\mathbf{x}'(A\mathbf{y})$, it can be interpreted as the result of applying a linear transformation to the vector $A\mathbf{y}$:

$$|\mathbf{x}'A\mathbf{y}| \equiv \|\mathbf{x}'(A\mathbf{y})\| \leqslant \|\mathbf{x}'\| \, \|A\mathbf{y}\|$$

Now taking the case of the linear transformation represented by $A\mathbf{y}$ we have,

$$\|A\mathbf{y}\| \leqslant \|A\| \, \|\mathbf{y}\|.$$

Thus, in general,

$$|\mathbf{x}'A\mathbf{y}| \leqslant \|\mathbf{x}'\| \, \|A\| \, \|\mathbf{y}\|.$$

8.5 Convergence of Matrices

8.5.1

If A denotes any matrix norm then the convergence of a sequence of matrices in the sense of this norm is defined as follows:

$$\lim_{k\to\infty} A^{(k)} = A \quad \text{if and only if} \quad \lim_{k\to\infty} \|A - A^{(k)}\| = 0.$$

Clearly, if $\lim_{k\to\infty} A^{(k)} = A$ then $\lim_{k\to\infty} \|A^{(k)}\| = \|A\|$.

8.5.2 Lemma 1

If $\|A\| < 1$, *then* $\lim_{k\to\infty} A^k = 0$.

Proof

$$\|A^k\| = \|AA^{k-1}\| \leqslant \|A\| \; \|A^{k-1}\|$$
$$\leqslant \|A\|^2 \, \|A^{k-2}\| \leqslant \ldots \leqslant \|A\|^k, \text{ and the result follows.}$$

Lemma 2

If λ is any eigenvalue of A then $|\lambda| \leqslant \|A\|$.

Proof

$A\mathbf{x} = \lambda\mathbf{x}$, for some non-null vector $\mathbf{x}$. Hence, for any mutually consistent vector and matrix norms,

$$|\lambda| \; \|\mathbf{x}\| = \|\lambda\mathbf{x}\| = \|A\mathbf{x}\| \leqslant \|A\| \; \|\mathbf{x}\|, \qquad \text{whence } |\lambda| \leqslant \|A\|.$$

Corollary

Since $\|A\|_2^2 \equiv$ maximum eigenvalue of $\bar{A}'A$ it follows that

$$\|A\|_2^2 \leqslant \|\bar{A}'A\|_1 \leqslant \|\bar{A}'\|_1 \; \|A\|_1 = \|A\|_\infty \; \|A\|_1$$

8.5.3 Theorem

In order that $\lim_{k\to\infty} A^k = 0$, *it is neccessary and sufficient that all eigenvalues λ_i of the matrix A should be less than 1 in modulus.*

Proof

Sufficiency follows at once from Lemmas (1) and (2) above. The proof for the necessity can be constructed along the following lines:

When a matrix A has n distinct eigenvalues, there always exists a similarity transformation which diagonalises A,

$$A \to H^{-1}AH = \text{diag}\,(\lambda_i).$$

If A has multiple eigenvalues then while it may not be possible to find a similarity transformation which reduces it to diagonal form, it is always possible to reduce it to the so-called *Jordan canonical form* (which is, in a sense, the "most nearly diagonal form"). First, we factorise the characteristic polynomial of A:

$$p(\lambda) = \prod_{i=1}^{k} (\lambda - \lambda_i)^{m_i},$$

where m_i is the multiplicity of the eigenvalue λ_i. A non-singular matrix M can then be found such that,

$$J \equiv M^{-1}AM = \begin{pmatrix} \boxed{J_1} & & & \\ & \boxed{J_2} & & \\ & & \ddots & \\ & & & \boxed{J_k} \end{pmatrix}$$

Here each of the blocks J_i is an $m_i \times m_i$ *Jordan submatrix* of the form,

$$J_i = \begin{pmatrix} \lambda_i & e_1 & 0 & \dots & 0 & 0 \\ 0 & \lambda_i & e_2 & \dots & 0 & 0 \\ \dots & \dots & \dots & \dots & \dots & \dots \\ 0 & 0 & 0 & \dots & \lambda_i & e_{m_i-1} \\ 0 & 0 & 0 & \dots & 0 & \lambda_i \end{pmatrix}$$

Each of the quantities $e_1, e_2, \dots, e_{m_i-1}$, appearing on the superdiagonal of J is either 0 or 1. Without loss of generality, we may assume in the argument which follows that all the quantities e_1, e_2, etc. take the value 1 in a typical submatrix J_i.

Now note that if $J = M^{-1}AM$ then

$$A = MJM^{-1}$$

and $A^k = MJ^kM^{-1}$.

To obtain the kth power, J^k, of the Jordan canonical form it is enough to examine the sub-matrices J_i. Typically we get:

$$J_i^2 = \begin{pmatrix} \lambda_i^2 & 2\lambda_i & 1 & 0 & \dots & 0 \\ 0 & \lambda_i^2 & 2\lambda_i & 1 & \dots & 0 \\ 0 & 0 & \lambda_i^2 & 2\lambda_i & \dots & 0 \\ \dots & \dots & \dots & \dots & \dots & \dots \end{pmatrix}$$

$$J_i^3 = \begin{pmatrix} \lambda_i^3 & 3\lambda_i^2 & 3\lambda_i & 1 & 0 & \dots & 0 \\ 0 & \lambda_i^3 & 3\lambda_i^2 & 3\lambda_i & 1 & \dots & 0 \\ 0 & 0 & \lambda_i^3 & 3\lambda_i^2 & 3\lambda_i & \dots & 0 \\ \dots & \dots & \dots & \dots & \dots & \dots & \dots \end{pmatrix}$$

and so on. It is then clear that $J^k \to 0$ (and so that $A^k \to 0$) as $k \to \infty$ if and only if $|\lambda_i| < 1$ for all i.

8.5.4 Convergence of the Series $I + A + A^2 + \ldots$

(i) Theorem

A necessary and sufficient condition for the series $I + A + A^2 + A^3 + \ldots$ to converge is that $\lim_{k\to\infty} A^k = 0$.

Proof

The necessity is obvious. For sufficiency, note first that if $A^k \to 0$ as $k \to \infty$ then all eigenvalues of A must lie inside the unit circle. Hence, $I - A$ must be non-singular and so $(I - A)^{-1}$ exists.

Now consider the identity,

$$(I - A)(I + A + A^2 + \ldots + A^k) \equiv I - A^{k+1}.$$

Since $(I - A)^{-1}$ exists we may write,

$$\begin{aligned} I + A + A^2 + \ldots + A^k &= (I - A)^{-1}(I - A^{k+1}) \\ &= (I - A)^{-1} - (I - A)^{-1}A^{k+1} \end{aligned}$$

The second term on the R.H.S. certainly tends to 0 as k tends to ∞.

Hence,
$$I + A + A^2 + \ldots = (I - A)^{-1}.$$

Corollary

For any matrix norm such that $\|I\| = 1$ we have

$$\|(I + A)^{-1}\| \leqslant (I - \|A\|)^{-1}.$$

Proof

$$\begin{aligned} \|(I + A)^{-1}\| &= \|I - A + A^2 - A^3 + \ldots\| \\ &\leqslant \|I\| + \|A\| + \|A\|^2 + \|A\|^3 + \ldots \\ &= \frac{1}{1 - \|A\|}, \quad \text{if} \quad \|A\| < 1 \quad \text{and} \quad \|I\| = 1. \end{aligned}$$

(ii) Theorem

Each of the following conditions is ***sufficient*** *for the convergence of the series $I + A + A^2 + A^3 + \ldots$*

(a) $\|A\| < 1$ *for any norm of A,*
(b) *There exists some matrix B, which is similar to A, such that $\|B\| < 1$, for some norm.*

Proof

(a) If $\|A\| < 1$ then for every eigenvalue λ_i of A we must have $|\lambda_i| < 1$. Thus, $A^k \to 0$ as $k \to \infty$ and so the series converges.

(b) If $A = HBH^{-1}$ then,

$$I + A + A^2 + \ldots + A^k = H(I + B + B^2 + \ldots + B^k)H^{-1}$$

and this converges if $\|B\| < 1$ for some norm.

(iii) **Theorem**

If $\|A\| < 1$, *then*

$$\|(I - A)^{-1} - (I + A + \ldots + A^k)\| \leqslant \frac{\|A\|^{k+1}}{1 - \|A\|}$$

Proof

$$(I - A)^{-1} - (I + A + \ldots + A^k) = A^{k+1} + A^{k+2} + \ldots$$

which implies that,

$$\|(I - A)^{-1} - (I + A + \ldots + A^k)\| \leqslant \|A\|^{k+1} + \|A\|^{k+2} + \ldots$$
$$= \frac{\|A\|^{k+1}}{1 - \|A\|}$$

8.6 Convergence of Iterative Processes

In Section 4.3.2, the general problem of iterative processes of the form

$$\mathbf{x}^{(r+1)} = M\mathbf{x}^{(r)} + \mathbf{c}$$

was discussed. Here, M is an $n \times n$ matrix and $\mathbf{c}$ an $n \times 1$ column vector. Starting from a certain initial approximation $\mathbf{x}^{(0)}$ we obtain successive approximations $\mathbf{x}^{(r)}$ which can be expressed explicitly in terms of $\mathbf{x}^{(0)}$ as follows:

$$\mathbf{x}^{(r)} = M\mathbf{x}^{(r-1)} + \mathbf{c} = M^2\mathbf{x}^{(r-2)} + (I + M)\mathbf{c} = \ldots$$
$$= M^r\mathbf{x}^{(0)} + (I + M + M^2 + \ldots + M^{r-1})\mathbf{c}.$$

If the process converges at all, say if $\mathbf{x}^{(r)} \to \mathbf{x}^*$, then it must converge to a solution of the equation

$$\mathbf{x}^* = M\mathbf{x}^* + \mathbf{c}.$$

8.6.2 Theorem

A necessary and sufficient condition for the iterative process

$$\mathbf{x}^{(r)} = M\mathbf{x}^{(r-1)} + \mathbf{c}$$

to converge for any initial vector $\mathbf{x}^{(0)}$ *is that all the eigenvalues of M should be less than 1 in modulus.*

Proof

For sufficiency note that the condition implies that $M^r \to 0$ and therefore that the series $I + M + M^2 + \ldots$ converges to the limit $(I - M)^{-1}$.

For necessity, suppose that $\mathbf{x}^{(k)} \to \mathbf{x}^*$. Then,

$$\mathbf{x}^* - \mathbf{x}^{(r)} = M(\mathbf{x}^* - \mathbf{x}^{(r-1)}) = \ldots = M^r(\mathbf{x}^* - \mathbf{x}^{(0)}).$$

This implies that

$$M^r(\mathbf{x}^* - \mathbf{x}^{(0)}) \to 0$$

and since this holds for any $\mathbf{x}^{(0)}$ we must have $M^r \to 0$. Hence, all eigenvalues of M must be less than 1 in absolute value.

Corollary

A sufficient condition for convergence of the process is that any norm of the matrix M shall be less than 1.

INDEX